# THE BRAIN THAT DESTROYS ITSELF

## THE EVOLUTIONARY ORIGIN OF NEURODEGENERATIVE DISEASES ALZHEIMER'S, PARKINSON'S AND MANY OTHERS

### KEITH C. M. GLEGG

IUNIVERSE, INC.
NEW YORK   BLOOMINGTON

The Brain That Destroys Itself
The Evolutionary Origin of Neurodegenerative Diseases
Alzheimer's, Parkinson's and Many Others

*The views expressed in this work are solely those of the author and do not necessarily reflect the views of the publisher, and the publisher hereby disclaims any responsibility for them.*

*iUniverse books may be ordered through booksellers or by contacting:*

*iUniverse*
*1663 Liberty Drive*
*Bloomington, IN 47403*
*www.iuniverse.com*
*1-800-Authors (1-800-288-4677)*

*Because of the dynamic nature of the Internet, any Web addresses or links contained in this book may have changed since publication and may no longer be valid.*

*ISBN: 978-1-4502-6328-3 (sc)*
*ISBN: 978-1-4502-6329-0 (dj)*
*ISBN: 978-1-4502-6330-6 (ebk)*

*Printed in the United States of America*

*iUniverse rev. date: 10/26/2010*

To The Caregiver

# CONTENTS

# INTRODUCTION

## IS THERE REALLY A BRAIN THAT DESTROYS ITSELF?

Yes, there really *is* a brain that destroys itself. And it's a brain just like the one in your head, and the one in mine. But not every one of the brains like ours will ultimately destroy itself. However, if the destroying begins, then the signs of this are described in what neurologists take to be the *symptoms* of "neurodegenerative diseases."

## NEURODEGENERATIVE DISEASES

Neurodegenerative diseases are among the most serious problems of modern neurology. They make up a large family of diseases that involve disorders of the central nervous system, mainly the brain and its neural attachments, hence the prefix "neuro." The diseases are "degenerative" because, once started, they generally continue to worsen. They entail various kinds of physical and mental impairment, and generally prove fatal. Those most often encountered are Alzheimer's and Parkinson's disease, but there are many others—surprisingly many, as I'll be showing. Their hereditary, genetically-associated nature has been largely confirmed.

## WHAT IS CONFOUNDING ABOUT THEM

Now, since neurodegenerative diseases involve the brain destroying itself, any evolutionary explanation of them, like this one, must involve the evolution of the brain. And this will have to explain how such evolution could have arrived at the *hereditary* nature of a self-destroying condition in a brain.

But, more than this, the explanation will have to address another feature of neurodegenerative diseases, which is that they are not infections, due to viruses or bacteria; they are not the result of cancers, due to errant genes. Instead, they begin and develop, so it appears, *spontaneously* or "naturally," as if, somehow, they "belong." And this leads to what is *confounding* about them: In spite of their being *hereditary*, and in spite of their appearing to occur *spontaneously* or "naturally," as if they "belong," no one has ever shown how they *could possibly* have helped members of the human or any other species *survive*—even with all their dark spontaneity, and their seeming as if they "belong," neurodegenerative diseases have continued to *appear* as having had *no survival value* for members of any species.

## SOMETHING MUST BE MISSING

It hardly needs mentioning that a brain's *hereditary* destroying of itself *spontaneously* or "naturally" is an *extremely significant* part of what it can accomplish. So, that its spontaneous auto-destroying should *now* be seen as just "disease," having had *no survival value*, suggests that something must be *missing* in our grasp of the brain's evolutionary history.

Indeed, what this suggests is that there must have been a time, in the evolutionary history of brain, when, *very surprising as I know it will seem*, what is *now* seen as merely spontaneous destructive "diseases" actually *did* have survival value, with their *loss* of survival value coming only *after* evolution added *something else*, that led to how we view the "diseases" *now*. And as we will see, the "something else" that evolution added was the *hereditary* constraint that *a normal brain **must sleep***. Remarkably, it is seeing when and how that happened which will allow the present explanation to account for *where* neurodegenerative diseases, in their *entirety*, came from, and *why* they are now the disparate and generally fatal disordering of behavior and life they are.

So, it is how all that transpired which makes up the greater part of the book's explanation—a surprising and revealing recounting (even for me!) of what's been missing in the evolutionary story of brain, and its neurodegenerative diseases—in a kind of evolutionary detective story.

In summary, then, the explanation presented here brings the *entire large family* of disparate neurodegenerative diseases—larger than ever before described—within a unique evolutionary framework. In doing so, it provides a *single explanation* which accounts for both the brain's degenerative auto-destructiveness which sustains the diseases, and the perplexing connection of the diseases to not just sleep but, more significantly, *disorders* in sleep. And, so far as I'm aware, this is the first such evolutionary explanation of neurodegenerative diseases.

## HOW THE BOOK IS ORGANIZED

In many respects, the present book is an outgrowth of my earlier one: *The Evolutionary Origin of Human Behavior.* And that shouldn't be surprising, since the object *there* was to trace the development of human behavior as this reflected the evolutionary development of the human brain, while the related object *here* is to show how that very same brain, still as evolutionary development, came to be one that, not infrequently, will destroy itself.

That has led to two chapters in the present book: Chapters I and VIII, being repeats of chapters in the earlier book, but with substantial additions and deletions, which help focus the story of the evolutionary development of the human brain on neurodegenerative disease. So this is not a second edition. Instead, unlike the earlier one, this book is pointing at an evolutionary property of the human brain that could very well be the source of as much continuing grief and challenge—for all the increasingly many involved with it, privately and governmentally—as any other aspect of the modern human condition.

## LANGUAGE, STYLE AND REFERENCES

The language, style and references are intended to make the book convenient for non-specialists familiar with Google in particular. Indeed, the existence of Google has allowed me to reduce, to the barest minimum, the references that are either cited or quoted in the text. So, for instance, when I say or imply that something is "known" without identifying a source, just insert some related words in Google, then choose from the usually many relevant topics suggested there, and sources galore will appear. Oh, academia, yes, I know you were never like that!

In a somewhat similar vein, I have used all the various forms of punctuation and emphasis, now so easily available with just a few keystrokes—italics, boldface, ellipsis (…), dashes, exclamation signs, quotation marks, brackets of all kinds—to give the impression that I'm *reading* the text **aloud**, as, indeed, reading was done at its beginning. So, the frequent emphases are not simply poor writing, they're in addition to that—reminders of what the reading rooms of early libraries must have sounded like!

My hope is that the book might help the rapidly growing numbers afflicted or otherwise involved with these diseases to see where they came from, why they tend to grow into one another, and why they are so ultimately trying.

But although the book is slanted toward the non-specialist, it should be of interest to a wide variety of specialists as well. Perhaps it will even allow them to at least wonder whether what they do *not* **now** recognize as neurodegenerative disease could be simply *one more* of its many faces.

In all my looking at neurodegenerative disease I have seen nothing funny—nothing at all. But this has not denied me a few moments of light relief, expressed here and there as humor. So, let not a passing smile mask my deepest feelings for the victims of neurodegenerative disease, and my undying amazement at the devotion of the caregiver.

# CHAPTER I
# EVOLUTIONARY
# EXPLANATION

## EXPLANATION

All we need do is look around us to be convinced that humanity and its works are special. Indeed, so compelling is the evidence that we might easily come to believe that we are "fully detached" from the rest of the living world. But, beginning some two-hundred years ago, attention began to be directed toward even more compelling evidence of our *attachment to,* rather than our *detachment from* the world of the living, as it stretches back and converges toward some, as yet, unknown precursor. What the evidence suggests, as it continues to accumulate, is that we, like all other members of the living world which exist and ever have existed, evolved, generation after generation, from previous forms.

But there are gaps in the evidence, some of which can be filled easily, and some less so. One of the latter relates to the subject of this book—the possibility that a "naturally" occurring neurodegenerative "disease" could have had *survival value,* even if *now* it appears not to. Is it possible to explain away such gaps simply as natural developments in a long process of evolution? In trying to see what the answer might be, it is useful to begin by looking at some of what contributes to valuable and tenable explanation.

Above all else, really valuable explanation must bring together a number of what had previously been taken to be *un*related phenomena. The reason for this is simply that any *particular* phenomenon, *by itself,* is usually "explainable" in a number—perhaps an unlimited number—of *different* ways. So it is only on expanding the range of the "separate" phenomena explained by a *single* explanation that we begin to reduce the number of possible explanations, and demonstrate the power of any particular one.

Of course, no imaginable explanation will explain everything, but it is reasonable to expect of any explanation that we should not be able to present it with some phenomenon which would require it to contradict itself in order to sustain explanation.

Another requirement that might reasonably be placed on any explanation is that it be so constructed as to be "falsifiable." What this means is that a tenable explanation will lead to some conclusion which a demonstration or experiment can either support or not support. Only explanations that have the capacity to bring on themselves and *survive* this kind of filtering by demonstration can be extended, with any assurance, to still further explanation.

Looked at in this context, evolutionary explanation, both for what it has explained and how it achieves explanation, appears as one of the great triumphs of the human brain. For, in one grand sweep, it tells us of a relation between all the biological phenomena that are to be found in every living entity that has ever existed and which will ever exist, at least on Earth. One of the ways in which it accomplishes this is by bringing sharply into focus the issue of survival.

## SURVIVAL

For any living thing to have a continuing presence in the evolutionary world, that is, to "survive," it must be able to meet challenges to both its physical and its physiological integrity.

By "physical integrity" I mean the ability of the creature to hold together physically. It needs to be able to resist being arbitrarily torn apart by external occurrences such as small changes in temperature, currents in water or air, or the destructive efforts of predators.

By "physiological integrity" I mean its capacity to resist simply falling apart from starvation. This entails primarily ensuring sufficient intake of food and adequate means for getting rid of waste.

## REPRODUCTION AND EXTINCTION

But, in order that a creature might have lasting evolutionary significance, something else needs to happen beyond this self-directed survival activity, which is going on continually between the creature itself and its immediate environment. Out of this self-directed activity in the here-and-now must come some reproductive process which yields a reasonably faithful replica of the creature, so as to form a next generation. And since this concerns evolutionary explanation, I need to remind myself, at least, that such explanation cannot include the notion that creatures act in ways, non-self-directed ways, which favor reproduction as such. The nature of the evolutionary process requires that a creature act simply to benefit itself, that is, act in its self-interest. The interactions between the creature and its environment then engender and support reproductive processes or they don't.

Evidently, not every creature will continue to enjoy a relation with its environment that supports continued reproduction. In the case of those that do, we say the environment "favors" the creature, by a process of "natural selection" (to use Charles Darwin's tellingly insightful phrase), for as long as this continues. When the efforts of the creature to sustain its own integrity no longer yield reproductive processes as well, there is finally no next creature, and the line of creatures, going back some number of generations from the last, is said to have become "extinct." This is simply another outcome of the relation between the creature and its environment, also mediated by natural selection.

## SPECIES

No process of reproduction or copying will produce perfect replicas. In particular, the process underlying replicas of living creatures is subject to small errors, mutations, which tend to persist in successive replicas. These mutations can, on occasion, alter the relation between the creature and its environment in such a way as to alter the reproductive process itself. When the reproductive process is favorably affected, we say the mutation is favored by natural selection. Another way of saying the same thing is that the "fitness" of the creature is increased by the mutation.

Evidently, the combination of a more or less steady stream of mutations with the endless filtering provided by the environment in the form of natural selection would lead to the emergence, in successive generations, of particular groups of creatures which enjoy increasingly special relationships to the environments in which they live. When such a group becomes so specialized and distinct from those around it that either no mating outside the group occurs, or such mating, if it occurs, no longer leads to successive generations, we refer to the new distinct group as a "species."

The process that leads to such a complete separating-out can hardly be very rare, since there are estimated to be between ten and twenty million existing species. Humans constitute one of these species, and my object in this book is to explain one of the processes that have led to *our* separating out as one of the species with a brain that, in some of its members, destroys itself.

## GENES

As it happened, it was possible for the remarkably brilliant Charles Darwin to assemble the first comprehensive statement of the nature of biological evolution without an awareness of what has since become its cornerstone. Darwin's observations and those of others, on both plants and animals, on which he based his revolutionary conclusions about the nature of evolution, were almost all made at or near the level of the unaided human senses. What has become clear, however, since his early efforts in about 1840, is that much of what he concluded from observations at this large scale can be traced to the workings of sub-microscopic chemical groupings that are known as "genes," these workings being referred to as the "genetic" process.

It is now known that the form and much of the behavior of all living entities are determined by the action of individual genes or groups of genes, and that these form parts of long chemical chains in which they are held together in patterns that are characteristic of the particular living entities of which they form a part. It might seem surprising that massive creatures like elephants and whales, and even ourselves, should have all their form and much of their behavior determined by such sub-microscopic strings of genes, but this is now well known to be so. It is therefore not uncommon to see references to genes being "expressed" in various features of living entities.

It is these genes, then, and the long chemical chains that link them, which carry the information that allows one generation of a species to produce offspring of the same species. So when reference is made to "mutations," it is within genes that the chemical mutations occur which account for the changes that can take place within species, from one generation to the next. And it is these genetic mutations, as they continue to occur, while their effects continue to be filtered by natural selection, which lead eventually to new species.

Evidently, the chemical assemblies that make up genes must be relatively stable in order to convey, from one generation of a species to the next, essentially the same information about its members. It is therefore reasonable to wonder how this stability can be reconciled with the mutations that seem to be necessary to give rise to new species. One answer to this (but not the only one) is that, even though genes are stable in the presence of the most frequently occurring forces which would tend to disrupt them, there is, at the surface of the earth, a steady flow of particles which are even smaller than genes, that originate both inside and outside the earth, and which have sufficient energy to easily disrupt a gene with which they might collide.

Although such collisions might occur infrequently in the life of any particular creature, they will occur more frequently in the species as a whole, and it requires only one member of a species to suffer a genetic mutation in order to set in motion a possible movement toward a new species. It is therefore reasonable to suppose that there is a steady stream of creatures in which mutations take place. Natural selection will suppress most of the mutations as expressed in the creatures which bear them, but some of the creatures carrying mutations will survive. As the process continues, new species will eventually appear. In this way, the necessary stability of genes is overcome infrequently but steadily, thus giving rise to the whole progression that we call evolution.

This, as many readers will know, is hardly even a sketch of what genes are, not to mention how they perform their marvels. But it will suffice for what I am setting out to do, for much the same reason that Darwin was able to do without genes altogether. The reason, of course, is what one might call the "level of disaggregation" of living creatures at which the discussion will proceed. I shall be proceeding mainly at the level of the whole creature, and two levels down, such as a brain and parts of a brain. As you will see, when proceeding in this way, it is convenient to be able to refer to genes at about the level at which they have just been described, but going any further here with their description would contribute little to what I am setting out to do.

## INSTINCTIVE BEHAVIOR

Since the young of many creatures emerge from eggs that hatch in isolation from their parents, and, in spite of this total isolation at birth, immediately execute behavior typical of their species, it is clear that such behavior must be part of the development of the creature that is determined by its genes. Such behavior is what we usually refer to as due to "instinct," that is, as "instinctive" behavior. Since instinctive behavior is determined by genes, such behavior must be passed on from parents to children, that is, must be "hereditary." However, in view of the possibility of a mutation occurring in a parent which is not expressed in that parent itself, but only in its children, "hereditary" behavior can only imply similarity, with some room for shifts due to parental mutations.

Of course, it is not only behavior that is executed immediately after birth that is instinctive, but it becomes increasingly difficult to identify, with assurance, which segments of the creature's behavior are instinctive and which are not, as a creature has a chance to mix with other members of its species, especially older members, and its behavior becomes more and more complex. Indeed, one might wonder whether there are *any* segments of behavior that are *not* instinctive.

A simple way of addressing this is to go directly to behavior in ourselves, in which we find components that are *not* hereditary. One such is language, since we quite evidently do not have to speak the same language(s) that our parents speak. If there is a hereditary component in language (and there is) it certainly does not control language at the level of the "dictionary" we use—the child of ancestors who speak only Chinese might speak only French. So there is clearly, in us, a vast domain of behavior that is not hereditary, and that is acquired from contact with, and by *copying* the behavior of other members of our species, which might, but need not include ancestors.

Of course, we also carry out instinctive behavior, since we, like all other mammals, do not need to be shown how to suck, for instance, just after birth. It is therefore clear that some process of evolution must have occurred that would lead to our incorporating, *in the same brain*, hereditary programs that control *instinctive* behavior, on the one hand,

and hereditary programs that allow us to *acquire* other behaviors by *copying*, on the other.

## COPYING

It is important to notice that whether we refer to acquiring programs of instinctive behavior by means of *genes,* or of *non*-instinctive behavior by *copying as imitation,* we are, in *both* cases, witnessing *the power of copying,* since the genetic process is, beyond everything else, *copying.* Indeed, if one had to identify the most basic process at work in biology, it would quite properly be *copying.*

## STRESS

We can come on a different way of viewing the evolutionary process by noticing that every creature, to survive, must be able to perform behavioral routines that allow it to continue satisfying its needs in the presence of a changing environment. But, although this must generally be the case, there will inevitably be situations in which, although a creature survives, there exists a distinct gap between the behavior that it *would have had* to execute in order to satisfy its needs *fully* and that which all the behavioral routines available to it *allow.* This gap, between behavior that would meet needs *fully* and what can *actually* be performed, sets up in a creature the condition we call "stress." Thus, stress is simply a reflection of the inevitable lack of a "perfect fit" between the behavioral endowment of members of a species and their environment.

But even if, over long periods in a creature's life, there could be a "perfect fit" in some average sense, there would still be short periods of heightened stress, such as those associated with recurring attacks of hunger, since there are no behavioral routines that can satisfy hunger *permanently.* Consequently, as the intensity of stress rises and falls, it can be viewed as "driving" much of a creature's short-term behavior, and as being able to change its short-term biological state.

It is therefore reasonable to say that surviving species are those that have evolved so as to limit the stress that their members experience, and so we can expect to find a variety of ways in which mutations and natural selection have combined to limit or eliminate stress in them. As we will see, *sleep* is one of these eliminators of stress, and it will serve to show how intricate a process for eliminating stress can become.

Of course, every surviving species is exposed to the possibility of extinction due to changes in its environment which would subject its members to a new and sufficiently severe condition of stress that cannot be reduced or eliminated by either existing behavioral routines or new routines that can be acquired soon enough to avoid extinction. In a related way, mutations in individual members of a species can lead either to the extinction of these members due to increased stress, or to conditions of reduced stress. In the latter case, the members and their offspring survive, and could ultimately lead, by the extension of such a process of mutation combined with the filtering of natural selection, to a new species in which there is a new kind of "non-perfect fit" between its behavioral endowment and its environment.

## STEPS IN EVOLUTION

The next point I shall raise relates to the necessarily small steps by which, in my view, evolution must proceed. That it should proceed in steps at all is clear enough since, being a biological process, it must depend, ultimately, on differences between various chemical substances. But, as we know, various chemical substances differ from one another in steps, simply because the parts that make them up are discrete atoms drawn from a limited set of about a hundred different elements. For there to be no steps, the chemical substances that form genes would have to be made up of parts drawn from some kind of continuous soup of possibilities; but, so far as we know, the world is not constituted in this way. So steps we shall always have. But why should there be "small" steps?

Here we don't seem to have the same kind of absolute necessity as with just steps, but the situation is heavily weighted against large steps. This is so because *many* factors must come together to favor the persistence of any variety of creature, generation after generation; so even the smallest disturbance in some capacity to relate effectively to its environment places the survival of the line of creatures at risk. Evidently, the risk increases with the size of the disturbance, and natural selection will, except in the rarest of cases, simply lead to the extinction of those creatures that are the carriers of large steps. And so, when I refer to "small" steps, I mean steps so small between one generation and the next that only very careful "comparison" of successive generations by the "environment" would show that there had been any step at all.

This poses a very fundamental challenge to evolutionary explanation generally since, in trying to follow some line of development, there will be no grand events of change to which one can point. The best one will be able to do is *imagine* some small step which, if amplified by a subsequent process of mutation and natural selection, would, after many generations, produce a large, clearly recognizable step, as would be apparent to the "environment," or, in the case where we are the recognizing part of the environment, apparent to our *unaided* senses. Many of what must be such amplified expressions of small evolutionary steps become evident by comparing the behavior of creatures that preceded us with our own, and so I shall be making such comparisons quite frequently, as has already been done in the foregoing section on instinctive behavior.

You might know that the question of big steps or small steps in evolution tends to rage as a great controversy from time to time. But I shall leave the foregoing as all there will be of that here.

## ABSENCE OF PRECURSORS

If an animal has a normal life-span of five years and needs six months from birth before it can have young, we can expect to have about ten generations living at any one time. However, given the small steps in which evolution must proceed, it might have required thousands of generations for the animal to evolve from an identifiably different precursor to what it now is. This suggests that we can expect such identifiably *different* precursors to be *absent* from the population of animals that we see around us. Thus, when we say, for instance, that birds evolved from reptiles, this should not be taken to imply that any living reptile is, or even looks like, the reptilian precursor of birds.

## ORDER IN GROWING

We are accustomed to seeing, in the usual growth of an infant, what is taken to be its "normal" course of development. But although it isn't so easy to see, there is also a normal course of development in a human embryo. Indeed, generally, there is a normal course of development for every kind of creature—just a single normal way of growing, particular to its kind. What, then, determines each of these particular ways of growing? Each of these particular ways of growing is determined by the *evolutionary history* of the creature. And it could hardly be any different, since evolutionary change can only occur in what already exists as previous evolutionary outcome; so, as a creature has come to being what it is, there is only one particular path along which it must have evolved—it can have had only a *single, unique* evolutionary history. And, since genes determine what normal growing will be, it is in genes that the unique evolutionary history of the creature is stored, and out of which comes the expression of this unique history.

Looked at more mechanically, there is only *one way* that the creature can come together so that all the parts can be *sure* to fit—the way its predecessors went together when all the parts actually *did* fit. In evolution, all parts came after others, with natural selection filtering the order, mutation after mutation. And this has imposed on growing that some things *must* come after others, in an order that can change only at the risk of all the parts not *quite* going together, with natural selection vetting the ensuing debacle.

## VESTIGES

To see what "vestiges" are we can start with mammals that live in the sea. There are many such mammals, of which seals are a good example. Being mammals, they did not originate in the sea. Instead, their ancestors had lived on land, and only later evolved into what they are now: mammals living in the sea. But, although seals now have *no legs*, when the *ancestors* of seals lived on land *they* must have had legs, as all land-living mammals must. So seals, mammals that they are, must have *lost their legs* somewhere along their evolutionary passage into the sea. Actually, the legs have not been completely lost, for there are traces of legs, "*vestiges*" of legs, buried under the skin of a seal. And this is merely one example of a common evolutionary process that can be summarized as follows:

> When some existing hereditary attribute of a species ceases to have *survival value*—like legs would in mammals that begin living in the sea, rather than on land—continuing mutations and natural selection lead eventually to a *new variety* of the species whose

members do *not* carry the excessive survival burden of a *then* useless attribute, as happened with what became legless sea-living mammals. If, as is generally the case, there remains a recognizable version of the attribute not *then* performing the function it did, before becoming *useless*, this is then said to be a "vestigial" attribute, and is called a "vestige."

## SETTING OUT TO GET SOMEWHERE

As you read along, it might appear, on occasion, that what's intended is that "evolution is setting out to get somewhere." Whenever this occurs, you can be sure that that is *not* what's intended, simply because evolution is never setting out to get anywhere. Indeed, as evolution proceeds, it's never setting out to get anywhere if only because it has no way of "knowing" what some as yet *unrevealed* somewhere might be! So, if, as you read along, you should come on evolution-setting-out-to-get-somewhere, treat it as merely showing how bad style can triumph over (otherwise?) good sense. Then use your own good sense and style to set things evolutionarily right.

## THE FORM OF ULTIMATE EXPLANATION

Is there, then, a form of ultimate explanation? I believe there is, as follows: Any explanation of a biological phenomenon that is not, or cannot be placed in an *evolutionary context* is sure, sooner or later, to prove untenable. In effect, the *ultimate* explanation of any biological phenomenon will be an *evolutionary explanation*. Indeed, it could be that the ultimate explanation of *any* phenomenon is an evolutionary explanation. That is why it is so significant that what you will now find is the first-ever evolutionary explanation of neurodegenerative disease, as such an explanation expands the current view of the range and nature of such disease.

# CHAPTER II
# THE EVOLUTIONARY
# EMERGENCE OF BRAIN

## BEFORE BRAIN

There were tens of thousands of varieties of creatures on Earth long before the emergence of organisms with a brain—hundreds of millions of years before. They included organisms that consisted of just *single* cells. Each single cell was enclosed in a thin membrane, part of whose function is simply to hold the interior parts together. Some of these are with us still, in the form of bacteria, for instance.

However, there were also other organisms that existed before the emergence of a brain, which consisted of *many* cells joined together. The cells in these organisms are held together by chemical substances that are a kind of flexible glue—proteins and sugars—the details of which needn't concern us here. And, as with the single-celled organisms, some of these are with us still, in the form of algae and yeasts, for instance.

But if such multi-celled organisms are to form more than just a clump of cells stuck together, and survive as some new kind of integrated, active organism, there must be a way for each cell to affect the condition of at least its adjacent cells. The way this is done is by one cell passing *chemical molecules* to its neighbors, out through its enclosing membrane, and in through its neighbors' adjacent membranes. Thus, in these early multi-celled organisms, the chemical condition existing in one cell could affect that of its neighbors by the actual transport of *molecules* directly to them.

And such organisms, "simple" though they be, are remarkable in their capacities for survival, since they are still with us, in what must be much like their original forms of hundreds of millions of years ago, albeit with mutations that must have occurred along the way. But if we now make a giant evolutionary leap, we can come on organisms that are still more remarkable, for their cells employ the previous transfer of *molecules* to achieve what is essentially *electrical* inter-cell *communication*.

Each cell that does this is stretched out like a long string, and an *electrical potential* (voltage) moves along its length if it is stimulated at one end by "something happening," say, an increase of pressure. This electrical potential is then used to energize some of its molecules which are transferred to an adjacent cell. But, now, it is the *fact* of there having been an increase in pressure—the *fact* that something had *actually happened*—which *initiated* the *electrical potential*, that is transferred between cells *via* molecules.

## CELLS THAT "AS IF" BETWEEN THEM

So, there's something *essentially new and very different* about this kind of transfer between these evolutionarily later-occurring kinds of stretched-out cells. For, although it is still *molecules* that are transferred between the first and second cells, *what* is being *conveyed* by *this* transfer is not molecules functioning *as just the molecules they are*, but, more importantly, it is molecules functioning as something *different* from what they are. They are no longer molecules involved in just *chemistry*: they are molecules involved in something *more* and *different*. For what is now happening is the passage between cells of a molecule that is not *itself* the "increase in pressure," but a molecule that is "**standing for**" it—a molecule that is not the "happening itself," but a molecule that *stands for* the happening itself.

And this is not just new, it's revolutionary. For the new kind of cell has come on a way of making a molecule behave, in going from cell to cell, "as if" *it* were the increase in pressure. So the new kind of cell has come on what, eons later in evolution's passage, would have some explorer drawing a "map" that would be "as if" it were the shape of some actual place; would have a teller of children's tales saying "once upon a time," as the warning that just an "as if" is about to unfold.

This is the emergence of the cells that, because they can, between them, "stand for" some happening, are the first glimpse, not of the impossible retaining of the happening itself, but, instead, the retaining of a "memory," that "stands for" what the happening had been.

This is the dawning, not of the impossible sending of happenings themselves, but the sending of *signals, messages,* from place to place, from one end of a cell to its neighbor, *telling* of their happenings. It is the first trace of cells that, together, will some day "pretend" there had been a happening—even if happening there had never been!

## CAJAL'S NEURON

These evolutionarily later-occurring kinds of stretched-out cells are what we know as "neurons," or "nerve" cells, or just as nerves. The understanding of the neuron, including how its electrical potential is generated and transferred between cells at a "synapse," took many years to be arrived at.

It reached essentially the way we now understand it in the remarkably original work of Santiago Ramón y Cajal, in the early twentieth century, for which he was awarded the Nobel Prize in 1906. He established the neuron as the primary functional unit of the central nervous system, making it the elementary item out of which the entire nervous system is assembled.

## HINTS OF BRAIN

The evolutionary development of the stretched-out neuron allowed the emergence of a sheltered location, housing a multi-celled *junction* of a group of neurons, with their extremities spread out beyond this "brain," as the sheltered junction would come to be known. And, because of the stretched-out nature of the neurons, able to move happening-bearing *messages* from place to place, this brain would eventually become involved in *central direction* of *parts* of the larger multi-celled organism that was its host, parts which were *remote from the junction,* remote from the brain itself. Indeed, a central brain could, using its stretched-out neurons, behave "as if" it were almost everywhere in its host, directing. But this kind of central "electrical" direction was such an unprecedented evolutionary development, that it must have been attended by special new necessities as it continued in any surviving host species. So let us look at what just one of these must have been.

We can begin by noticing that, if a brain is to direct remote parts of its host, then it will need some kind of catalog, some kind of "map" showing what the parts *are*, and *where* they happen to be, in their overall relation to the host. Furthermore, the brain will have to keep this map *updated*, so that when, at any time, it tries to control any particular part, it begins by "knowing" at least where the part happens to be at that time. Indeed, if central direction by a brain is not to become just growing chaos on its way to extinction, there needs to be more than just a "map"; the map will need to be *continually updated*. And there'll be more said about this continual updating shortly.

It will be convenient to have a name for this map, and I'm going to call it the "configuration store," since it is what has to carry the "configuration" of the host's parts. But so that I can identify the configuration store with something that actually came to exist in the brain, rather than in just my imagination—as just an "as if"—I'm now going to go to some of the evidence that such a store *actually exists* in a brain like ours, for instance. Following that, I'll go to the difficulties that must beset the workings of such a configuration store and the ultimate consequences of those difficulties for the emergence of ... neurodegenerative disease.

## EVIDENCE FOR A CONFIGURATION STORE

Between 1947 and 1950, Dr. Wilder Penfield did some most remarkable things. Working on a patient undergoing surgery for epilepsy, whose brain was exposed, and by stimulating various places in it, then seeing what moved, he was able to show that there was a *mapping* of "motor" parts of the body—hands, fingers, feet, tongue, and so on—*onto various locations in the brain.* He named this mapping the "motor homunculus."

But this remarkable and revolutionary demonstration led to a great deal of controversy, some of it distinctly negative, regarding what such a mapping really meant for the actual working of the brain. Much of this controversy can be found elsewhere, but is of little interest here. However, I must mention a remarkable letter, in a prestigious journal, in which the author dismissed the possibility of a map in the brain "standing for" anything really worthwhile—using just *words*, mind you, that must, themselves, have been presumably "standing for" something, in the writer's famous editorial brain. Oh, the joys of spending a life in reporting on the big, blue beetles in the bark of a birch, but how hard it is, then, to recognize, as a *forest*, some great expanse of birches, newly revealed by a rare genius!

What was, and still *is* of immense importance and interest is Wilder Penfield's showing that, present in the brain, there is a *mapping* of the motor parts of its host, the "configuration store" that there simply must be, if a brain is to effect central direction of these motor parts.

## KEEPING THE CONFIGURATION STORE UPDATED

But having just a "configuration store" would not have been enough for a brain to achieve central direction of the parts of a member of a host species that could escape extinction. Because, for central direction to amount to anything more than just increasing chaos and ultimate extinction, the configuration store would have had to be *continually updated*.

There's no mystery about this, and you can see why simply enough. So I'm going to tell you something you'll do *after closing your eyes* (in a moment I'll tell you why it's so important that you will have *closed your eyes*). Here's the something: Have your brain direct one hand to touch an ear, then to touch your nose, while the other hand is directed to touch your other ear, and then your tongue, *fast as you can*. All that something is what you should have done only *after closing your eyes*.

Evidently, because *your eyes were closed* while all this was going on, it must have been your *brain*, and your brain **alone**, in which the actual *results* of the direction were being kept track of, so that the hands could be directed to some *new* place, on and on. And, that all this could happen *with your eyes closed* is a striking reminder that the earliest brain emerged *before* the evolutionary emergence of eyes—a brain's eyes—that would come only later.

So, *closing our eyes*, then letting our brain go to work in its *fast* directing is a way of allowing ourselves to get a glimpse of what a brain must have been like even at its earliest emergence. And part of what we can catch in such a glimpse is that, for the brain to have survived as anything remotely like what we have now, it must, at its earliest dawning, have carried not only a configuration store of its hosts parts, but a *fast-acting dynamic configuration store*—one that is *continually and rapidly* updated. And it must have been only the new kind of *fast electrical* signaling that *neurons* brought with them, which could have provided a dynamic configuration store, allowing direction with anything like the *speed* that we can now so easily demonstrate.

## ERRORS IN THE CONFIGURATION STORE

There are lots of places, especially lately, in which there are man-made *fast-acting* stores, of various kinds of "happenings," based on *electrical* signalling. Every compact disc, every CD, is such a store of "happenings." And one of the features of all these stores is that *errors* occur in them—a few digits that shouldn't be there, a few that should be but aren't—even after the assemblers of such stores have tried as hard as they can to avoid them.

Indeed, the errors are such a fundamental part of these stores that a whole science of how to live with them has emerged. Some parts of the science are known as *error-correcting codes*. The result is that, when you put your CD in the player, what you *hear* is coming out just a little *later* than it's *coming off* the CD, the delay being the time for the player to apply an error-correcting code, so that the trumpet note won't sound off key, or too soft. But even if an error wouldn't have been all that disturbing in what a trumpet is supposed to sound like, it could bring on a very worrisome surprise if the store, with an un-corrected error, carried what your or my *bank account* was supposed to look like—up errors, evidently, less worrisome than down!

So, *errors* in *fast-acting stores* based on *electrical* signalling there have always been, and always will be. Which is telling us that *errors* there have *always* been in the *fast-acting* **configuration store,** assembled from **neuron-based electrical** signalling, **that a brain must have,** and errors there *always* will be.

Now, this might be sounding a lot more complicated than it needs to, with error-correcting codes, and what not. Therefore, since we're going to be making important use of an error-correcting code later on, let us ask the following question: What would we do if we wanted to send, to someone, a message that was as free of errors as we could reasonably make it, but over a communication channel that we *know* can commit errors? Well, we could appeal to one of the simplest of error-correcting codes, based primarily on *repetition*, as follows. So, say the message we wanted to send was this one:

*This message is not to contain any errors.*

We could send it *three* times, for example, to be received, errors and all:

*Thin message is now to convain any errors;*
*This messing is not to contain ant errors;*
*This message if not lo contain any ertors.*

Then, as the key to the code, we say that the message to be taken as the *final* message—the one that will be the *most free of errors*—is the one that is obtained by making up the message from *all* the words that *appear* the *same* way *at least* **two** times. And, as you can see, on comparing the three messages sent, that gives:

*This message is not to contain any errors.*

Evidently, the number of times the message needs to be repeated increases as the length of the message increases, and the communication channel becomes increasingly prone to error. So, for instance, if the message was what is on this page, with its 450 words (2,000 characters), it might become necessary to repeat the message hundreds of times, depending on how error-prone was the channel, and how free of errors we wanted the final message to be. What I now want to emphasize, out of this, is the following:

> An *error-correcting code can be constructed*, allowing arrival at a *message "almost" free of errors*, by *repeating* the *same* message with its errors *over and over*, and using, as the *final* message, *the version obtained by taking together all the parts that keep on repeating*. And this would apply to the parts of an *image*, if the parts were sent as a message.

All of which leads to a grand question in the evolution of the brain: What was the evolutionary path along which the inevitable errors in its configuration store led the emerging brain? And the answer is long enough, and fundamental enough, and revealing enough that I've set aside a whole new chapter for it.

# CHAPTER III
# THE ONE-SHOT BRAIN

## REFRESHER

*Errors—in fast-acting stores* based on *electrical* signalling—there have always been, and there always will be. Which is telling us that *errors— in* the *fast-acting* **configuration store,** assembled from **neuron-based electrical** signalling, **that a brain must have**—there have *always* been, and there *always* will be.

All of which leads to a grand question in the evolution of the brain: What was the evolutionary path along which the *inevitable errors* in its configuration store led the emerging brain? And the answer is long enough, and fundamental enough, and revealing enough that I've set aside this little chapter for its telling.

## ANSWER

We begin the answer with the earliest of brains directing parts of its host. As the directing continues, errors in the brain's configuration store begin to accumulate. At first, the errors are so small that there's no sign of them appearing in the behavior of the host's parts that are being directed. But, eventually, as the errors accumulate, their effects begin to be evident: a part trembles a bit, as the direction that would have put it in a certain place falters, moving it back and forth, about the directed place. The errors keep on accumulating, until the effect of the brain's now confused direction, on the behavior of the host's parts, endangers the host's survival. And this brings us to a Y-junction along the path of the brain's evolution.

Down one arm of the Y, the brain simply continues, unchecked, to accumulate errors. And although there might, indeed, have been an evolutionary excursion down this path, it would have led to a dead end, with members of the host species endowed with such a brain being driven to extinction, leaving no present-day offspring. So this possible arm of the Y is of no further interest.

But the situation is totally different down the second arm. For, down this arm, the intensifying *stress* in the host, due to the increasingly confused direction of its parts, brings on a *new brain-condition* (that I'll say more about in a moment) in which the now offending brain is **disabled**, by having the activity of its neurons *destroyed*. The result of this is that the host then continues to live without the benefit of a working brain, much as the host's *multi-celled evolutionary antecedents* had done, without having neurons at all. But the host would have had whatever survival benefit the brain's direction of its parts—even just the direction by a transient, one-shot brain—could have provided.

This is now a ***seminally new place*** in the *evolutionary development* of a brain's host. For now it is endowed with a brain which, when the *inevitable* **errors** in its configuration store reach a level that endangers the host's survival, the *stress* in the host leads to the brain **destroying itself**. And this is *seminally new*, because what we now have is a **host species** in which its members owe their **ongoing survival** to the **hereditary auto-destruction of the brain they each now carry**, whenever **errors** in the brain's **configuration store** entail behavior that **endangers** a member's survival.

## THE "NEW BRAIN-CONDITION"

As just promised, I now return to the intensifying *stress* in the host, due to the increasingly confused direction of its parts that brought on a **new brain-condition**, in which the then offending brain was *disabled*, by having the activity of its neurons *destroyed*. So, what might this "new brain-condition" have been?

The most likely "new brain-condition" would have involved merely a small change in some condition that was *already present* in the brain. This could consist of no more than an increase in the *normal concentration* of some substance ordinarily involved in the growth and development of neurons: substance X, say. So, the bringing on of the "new brain-condition" consists of *increasing the level of X* from its *normal concentration* in the brain to one that renders X **toxic** *to the very neurons* whose growth, at its *normal* concentration, X had actually *supported.*

That a substance which normally supports the growth of cells at a *certain* concentration can, at *increased* concentration, become *toxic* and deadly, is common knowledge to the gardener. For, giving the lie to "more is better," there's the fertilizer that *supports* growth of plant cells at one concentration, but which becomes *toxic* at increased concentration. Of course, there are no neurons among plant cells, but the same connection, between *increased concentration* of some normally growth-supporting substance and *toxicity*, applies to these different kinds of cells, as well.

Such a brain-process for auto-destroying the activity of its neurons, involving only increases in concentration of an *existing* substance, would seem much more plausible than any releasing of a *new* substance, by the host's stress. It provides the kind of "economy" in explanation that tends to recommend one explanation over another. Furthermore, there seems to be some growing research support for this process. So, in whatever follows, this will be the underlying assumption concerning how a brain auto-destroys its neurons, or, at the very least, eons ago, *began* auto-destroying them.

## ONE-SHOT BEHAVIOR IN NATURE

That a brain should emerge in which it acts just *once*, then "dies," might seem odd enough to be unlikely. But there are examples of such one-shot behavioral episodes in other places.

The commonest is probably the behavior of mayflies (*Hexagenia*). They carry out their mating *once*, following which the male dies quickly, and the female later, having laid her eggs. And there is another example, this time involving a chameleon in Madagascar (*Furcifer labordi*). Females lay their eggs in February, and the entire adult population has died by March.

So, that a brain, at its earliest emergence, should act just *once*, over a period of time, then die, is not a complete rarity as creatures have evolved. Let us, therefore, move on to what might bring us to a brain as we know it—not a one-shot brain, but, instead, one that continues to function, time after time, during the life of its host. That will follow a brief look at a body without a brain.

## A BODY WITHOUT A BRAIN

Let us go back a few paragraphs to where I say: "...the intensifying *stress* in the host, due to the increasingly confused direction of its parts, brings on a *new brain-condition* ... in which the now offending brain is *disabled*, by having the working of its neurons *destroyed*. **The result of this is that the host then continues to live without the benefit of a working brain, much as the host's *multi-celled evolutionary antecedents* had done, without having neurons at all.**" (I've added **boldface** to the earlier text.).

What this implies is that the host ends up as a body without a working brain. And since this might seem unusual, even unlikely, it is useful to notice that this is certainly not confined to merely the earliest host-with-a-brain, of the present chapter. For, in *us*, though evolutionary eons away, it is indeed possible for a brain to "die," from one cause or another, leaving a multi-cellular remainder-body to function as an essentially brainless host. What, in our situation, is significantly different from that of the algae-like multi-cellular brainless remainder of the present chapter is that, while that remainder can continue to feed and get rid of its wastes all by itself, we—the relatively complex feeders and waste eliminators we've become—are unable to do so in a brainless state.

But all this leads to a much more general suggestion, involving the traditionally six "vital organs": brain, heart, lungs, kidneys, liver and pancreas. For what it all suggests is that, among "vital organs," some are "more vital" than others. And the reason for this would be that, as we have just seen, an evolutionarily *later* organ, a brain, can "die" *without* the *total death of its host*; since the host then falls back to what is a viable evolutionarily *earlier* condition, one in which the *later* organ simply *had not yet existed*. So, evolutionarily *earlier* organs would, in a general way, be *"more vital"* than *later* ones. And this, in a curious way, would simply be a reversal of the path along which the host had presumably grown from "birth"—from one viable evolutionary state to the next such state, that had come only later.

Going a little further with this, one might reasonably surmise that, between the algae-like, multi-cellular brain-destroyed remainder of the present chapter, and evolutionarily much later humans, there were whole species of creatures whose members, following the **survival-**

related auto-destruction of their brain, carried on a normal life. Indeed, it could even be that there are such species still living, but in which this feature of their lives, having had no apparent importance, has gone un-reported.

Not really directly related to my main story, but interesting, is the possibility that the host of a brain carries within itself what, in a sense, are *two* parallel cellular systems: one system made up of neurons which move, between themselves, molecules that merely *stand for* things, involved in "as if's"; and a second system made up of cells which move, between themselves, molecules that function, not as signals of anything, but simply *as just the molecules they are,* involved in just chemistry. Indeed, the brain with its neuronal system would appear to be *almost* separate from the other system, in that, even if the brain made up of neurons should die, the rest of its host, consisting of cells that move molecules, as just the molecules they are, between themselves, can continue to function as a living organism, albeit unable to "as if"—functioning like algae and other ancient neuron-less, brainless multi-celled organisms did, and do.

This all suggests that the host of a brain carries within itself what, in a sense, are *two* parallel cellular systems: multi-neuronal-cellular brain, and multi-other-cellular body. And it could even be that the host of a brain carries within itself, as *two* parallel but necessarily connected cellular systems: a body *and* a "mind," which came only later, as evolution would have had it, capable of no less than the bewildering "as if" that's a dream.

Interesting as all this may be, let us now return to the main story, and see what we might learn about a brain that doesn't destroy itself in a singular episode, but which goes on to doing its directing, time after time.

# CHAPTER IV
# A BRAIN THAT DOES ITS
# DIRECTING TIME AFTER
# TIME

## WHERE TO LOOK

We are at a **seminally new place** in the *evolutionary development* of a brain's host. For now it is endowed with a brain which, when the **inevitable errors** in its *configuration store* reach a level at which brain-directed behavior endangers the host's survival, the **stress** in the host leads to the brain **destroying itself**.

So where should we look for the path that evolution *must* have followed in arriving at a brain, like ours, say, which, unlike the one in the previous chapter, does *not* do itself in every time the errors in its configuration store build up—a brain that can do its directing time after time, during the life of its host?

## FINDING THE ANSWER

Evidently, the answer will have to involve what must have been the evolutionary emergence of a host carrying something distinctly *new*. And, given the existence of a brain like ours, part of what simply *must* have been distinctly *new* was that a host emerged with *an augmented genetic endowment* expressed in a new kind of brain which *cleared the errors* in its configuration store **before** it would have gone on to its hereditary self destruction. But for that, you wouldn't be reading this.

So I'm now going to sketch how this new kind of brain would have carried out the clearing of errors, then have gone on to *repeating* the directing of its host's parts, *without* having destroyed itself. And it's a "sketch" because I want to go quickly, passing over a lot of details, some of which will be addressed later.

Picture, then, the behavior of a host in which the directing of its parts is being carried out by this new kind of brain. And, in this brain, as in every brain, errors accumulate in its configuration store. As the errors accumulate, the brain-directed behavior begins to endanger the host's survival. But the ensuing ***stress*** in the host ***does not*** lead to the brain destroying itself, *as would have occurred with the previous one-shot brain*. ***Instead***, the *augmented genetic development* of the host is now expressed in the *stress* releasing a burst of brain-substance (which I'll get to later) which induces a state of ***quiescence*** in the host—its motor parts become **still**.

During this state of ***quiescence***, the configuration store is ***not*** being continually refreshed with ***new, changing*** configurations from the host's *now **stilled*** motor parts. Instead, the configuration store is forced to ***repeatedly*** acquire the ***same stationary, resting configuration***—*w*hatever it happens to be—and becomes reloaded with **that *same stationary*** configuration, *over and over* again.

And now I recall an earlier conclusion, reached in Chapter II, regarding error-correcting codes:

> An *error-correcting code can be constructed*, allowing arrival at a *message "almost" free of errors*, by *repeating* the *same* message with its errors *over and over*, and using, as the *final* message, *the version obtained by taking together all the parts that keep on repeating*. And this would apply to the parts of an *image*, if the parts were sent as a message.

Returning attention to the configuration store, it continues to ***repeatedly*** acquire the ***same stationary, resting configuration***, whatever it happens to be, *while applying an error-correcting code*, based simply on *repetition*—*retaining* in the configuration store only those parts of the incoming image that **repeat**.

Then, slowly, the effect of the burst of quiescence-inducing brain-substance wears off, and the host recovers its active behavior, with its *configuration store* initially "almost" **free of errors.**

This new kind of *error-correcting* brain can now resume the directing of its host's parts. And it can do this, *time after time*, as its stress-induced error-clearing periods of quiescence come and go.

## A BRAIN MUST SLEEP

Although, in this way, the host's *augmented genetic development* would have succeeded in clearing the errors from a brain's configuration store, it is important to notice that it does so at the expense of what would be an ***endless series of periods of quiescence*** which interrupt the host's full functioning. These periods of quiescence, which characterize the functioning of a host with a brain having the new ability to clear the errors in its configuration store, are the earliest evolutionary manifestations of what we come to know as "**sleep**." When the creature is no longer in one of these periods, we say it is "awake," by which we evidently mean: functioning with its neuron-based brain doing its directing, with a configuration store operating as it did earlier—initially free of errors.

It is important to notice that there is no particular configuration-of-motor-extremities to which the configuration store must be returned during sleep—there is no "zero" configuration of motor extremities from which the waking state must always begin. The waking state begins with whatever essentially "fixed" configuration of motor extremities happens to have prevailed during the *stillness* of sleep, and goes on from there, by continued refreshing of the configuration store, with whatever stress-inducing errors accumulate in the course of refreshing. The stress induced by these accumulating errors, while awake, brings on the next period of sleep, in an endless series of sleep and waking.

## NAPS, SIESTAS AND LONGER SLEEPS

Well, if that's what sleep allows a brain to do to itself by applying an error-correcting code, then *how* **long** *should sleep last*? And it is indeed interesting that I should now be asking that question—heretofore, with good reason, never asked—with hope of finding any kind of answer **at all**!

The answer can start with a follow-up question: How long does it take for the host's motor parts to complete the refreshing of what must be a dynamic configuration store—for up-dating Penfield's motor homunculus? You can make an estimate of this as follows: *With eyes closed*, change the configuration of your motor parts by, for instance, repeatedly bringing together your first finger and thumb, *as fast as you can*, and notice that this can be done something like *ten times per second*. The configuration store must be being refreshed *at least* ten times per second, or six hundred (600) times per minute, to make this possible. But to keep the following piece of arithmetic simple, we'll say that, to keep the configuration store up-dated, it must be refreshed at least a thousand (1,000) times per minute.

We are now going to apply an error-correcting code like the one mentioned earlier, based on *repeating* a message known to contain errors. So let us assume (guess, if you prefer) that, to reduce the errors in the configuration store to a level that will allow renewed stress-free control of motor parts—to render the configurations store (the "map" in Penfield's motor homunculus) "almost" error-free—there are ten thousand (10,000) repetitions needed in the configuration-store version (image) of the *sleep*-**stilled** motor parts. Then the time taken for the clearing of errors in the configuration store would be:

10,000 repetitions divided by 1,000 repetitions per minute.

This gives **ten** (10) minutes—sleep would need to last some **ten minutes** to reduce the errors in the configuration store to a level that will allow renewed stress-free control of motor parts. If we allowed the estimate of the number of required repetitions to range between 1,000 and 100,000, sleep would need to last no more than between one minute and a hundred minutes—between a minute and less than two hours.

I am going to take the result for 20, 000 repetitions, which gives **twenty minutes** for the necessary length of sleep, because there is a great deal of evidence that this is about all that's needed to return the configuration store to a level that will allow a brain's renewed stress-free control of its host's motor parts.

The everyday evidence comes from what has come to be known as a "nap," the nice little bit of sleep—twenty minutes or so—that is all one needs to be "ready to go again"; and from the refreshing "doze" captured in a Spanish afternoon's "siesta." That is also roughly the length of time recommended, by authorities on driving automobiles safely, for the amount of time that one should go off the road and *sleep*, to avoid the dangers of failing motor control, due to a *need* for *sleep*.

All of which, as the foregoing rather crude estimates and calculations suggest, is saying the following:

> It takes something like **twenty minutes**, and 20,000 repetitions during **sleep**, coming from a host's **stilled** motor parts, for the errors in the configuration store of a brain like ours to be returned to a level that allows renewed rapid, precise, stress-free control of our motor parts. Presumably, that many repetitions are required because of the enormous size of the "messages" involved in making even a single "image" of the host's motor parts in the configuration store.

Of course, evolution never begun with a brain as massive as ours—with its hundred-billion (100,000,000,000) neurons—simply couldn't have. Indeed, there are still creatures that must have been here long before us, who have, not hundreds-of-thousands-of-millions of neurons, but just a few hundreds. So evolution's early attempts at a brain must have been truly modest, compared with where braininess has led to in us. From which we can see that evolution's early arrival at *sleep* must have involved, not much of an *hour*, but *less than a* **minute**, to get an early homunculus ready to renew it directing of "motor" parts, free enough of errors to make for survival, and a nod from natural selection.

But it is not any presumed "accuracy," achieved by the foregoing assumptions and calculations, which is of real interest. Rather, what **is** of real interest is the fact that the whole evolutionary process being advanced here, of a brain that must *sleep*—one with a configuration store that had its errors cleared by the evolutionary emergence of sleep—could lead to **any** kind of tenable estimate **at all**, of the *duration required for sleep*.

There's a question that comes easily out of the twenty or so minutes of sleep that seem necessary to bring the configuration store to a practically error-free state: What would be the likely result, in humans, of repeated short periods—ten minutes or so—of *interrupted* sleep? The answer would seem to be that, due to a growing, overlapping error-state in the configuration store, this would lead to a greater lack of motor control, for instance, than would no sleep at all, for, say, a few days. Perhaps, as torture, repeated sleep *interruption* is more behaviorally compromising than extended sleep *deprivation*.

It could be that the kind of sleep being discussed here should be referred to as "primal" sleep, for this is the part of sleep that must have been its earliest evolutionary form—the part that allowed the *original* transition, from the one-shot, to the life-long brain—and probably the form still evident as all the sleep there is in reptiles, for instance. It must also be the shortest recognizable stage in sleep, which, as further evolution would have it, was to be followed by later and longer periods of sleep, as in us. But the story of these other longer periods of sleep, including that of Rapid-Eye-Movement (REM) sleep in particular, belongs elsewhere.

## THE BRAINY FLATWORM

So when did the first "real" brain—the first brain to know *sleep*—tell some ancient creature how to move its motor parts? The fossil evidence suggests that the first such brain probably had an early variety of flatworm as its host. And this would mean that, as modern dating of the fossil evidence would have it, the first sleeping brain was already active some 560 million years, or some 30 million generations ago.

But the earliest evolutionary emergence of a brain must have been some many (millions? of) years before the appearance of the fully operational brain of the flatworm. All of which says that the processes of evolution have taken a very long time—indeed, a significant part of the time since Earth grew out of a celestial mist—to arrive at a brain as remarkable as ours. Not surprisingly, perhaps, it has taken a very long time to arrive at what might just be the most complex entity in all known existence.

## NOW THAT A BRAIN SLEEPS

Let us summarize the situation that evolution presents, now that a brain sleeps. What we have is a host whose brain carries **two hereditary** modes of behavior. The **earlier** mode is such that when the **inevitable errors** in the brain's *configuration store* reach a level at which brain-directed behavior endangers the host's survival, the *stress* in the host leads to the brain *destroying itself*.

The mode that evolved *later* is one in which, when the *inevitable errors* in the brain's *configuration store* reach a level at which brain-directed behavior endangers the host's survival, and the *stress* in the host *would have led* to the brain destroying itself, *instead,* the brain initiates *sleep,* during which the *errors* in its configuration store are *cleared,* and the brain *returns to directing its host's motor parts,* until accumulating errors bring on stress in the brain's host, then another period of *sleep,* and so on, *time after time.*

What had earlier been a one-shot brain has evolved into one that can resume its directing time after time, because the host's genetic endowment has now given it *sleep.* And what we can see from this is that, so long as there are no *disorders in sleep—quasi-absences* of sleep—the errors in the brain's configuration store will be *cleared,* and the hereditary mode of behavior in which it destroys itself will *not occur.* But this is, indeed, only *if* there *are* no disorders in sleep. So let us see if there *can* be disorders in sleep, and what will happen if there are.

# CHAPTER V
# DISORDERS IN SLEEP: ONSET
# OF NEURODEGENERATIVE
# DISEASE

## Hereditary and Accidental Disorders in Sleep

As explained earlier, sleep emerged as a *hereditary* feature of its host, expressing the host's augmented *genetic* endowment. But if sleep is such a feature, then there must be **genetic** *disorders* that are expressed in **sleep** *disorders*—there must be at least *some* sleep disorders that are *hereditary*. In effect, there must be some sleepers who suffer sleep disorders simply because one or both parents suffered sleep disorders, before them.

However, we can go further. For, clearly, not *all* sleep disorders will be expressions of *genetic* disorders, since all sorts of disturbances, external and internal to a brain, can cause sleep disorders—even just writing a book on neurodegenerative disease can cause sleep disorders! In effect, sleep disorders can be either *hereditary* or *accidental*.

Furthermore, there are other kinds of accident, other than accidental sleep disorders, which have an effect in a brain *similar* to that of a sleep disorder, because they are traumatic enough to produce "un-clearable errors" in its configuration store. Such accidents include stroke; drug intake; direct head injury; encephalitis; concussions as in some sports, as examples.

As we will see, these sleep disorders, hereditary or accidental, will have a lot to do with the onset of neurodegenerative diseases, and whether *they* are, or are not, hereditary. And, except in some special situations, I'll refer to them all as just "sleep disorders."

## DISORDERS IN SLEEP AND ONSET OF NEURODEGENERATIVE DISEASE

The new brain, which clears its configuration-store errors *when it sleeps*—the evolutionary successor to the one-shot brain—marvelous though it be, has now led to a potentially *destructive* situation which arises if its host should begin to develop *disorders in sleep*. Let us see why.

The potentially *destructive* situation begins to appear because the *clearing of errors* in the brain's configuration store **should** now occur during *normal* sleep. But this clearing of errors *will fail* to occur when there are **disorders in sleep**—*quasi-absences of sleep*—and this *failure* will result in **errors** *accumulating* in the brain's configuration store. However, when this *failure* **persists**, the brain—acting as though evolution *had never brought* **sleep**—will *revert* to what it did earlier, as a one-shot brain, *before it acquired sleep,* then begin applying the *hereditary* process in which *it progressively disables its own ability to function as a brain.* And this is now disastrously destructive for the host since, when this brain reverts to its *hereditary auto-disabling,* the brain it is disabling has *already assumed a vital, complex role in the survival of the host,* **because this brain now includes whatever normally induces sleep**, with what sleep then *normally* **should** do to *inhibit* the very auto-disabling that is now under way.

Consequently, and disastrously for the host, when there are *persistent sleep disorders*, and auto-disabling has **begun** in this new kind of brain that normally sleeps, then the **more** *it is driven* to auto-disable itself, **still more** is it driven to auto-disable itself—auto-disable the **neural** components that make it up—on and on, **degeneratively**. So, when such destructive brain auto-disabling commences, we are witnessing the onset of *a "neuro-degenerative disease,"* as we now call it—one which had been **preceded by persistent disorders in sleep**, and has its primary focus in the **degenerative auto-destruction of the neuronal components that make up a brain**.

Becoming evident, from all this, is the following:

There **is**, and always has been, *an evolutionary* ***connection*** between ***persistent disorders in sleep***— amounting to its quasi-absence—*and the development of a destructive* ***neurodegenerative disease***.

This would be simply a re-enactment of the *evolutionary history* of the *source* of such disease, with some percentage of the population carrying, to later offspring, their earlier genetic endowment which, *before* the evolutionary emergence of sleep, was expressed in—irony of confounding ironies—auto-disabling of a brain which *then* had ensured its host's ***survival***.

Given this, disorders in sleep could not be properly regarded as the *cause* of neurodegenerative disease, they only leave the door ajar for it to enter—they become the un-guarded portal beyond which the destructiveness of neurodegenerative disease can begin.

## IN HUMANS ONLY

We can see from the foregoing that, if there are no persistent sleep disorders—just *sleep*—then there will be *no* neuronal auto-destruction, and no neurodegenerative disease. Therefore, the *survival value* that *hereditary neuronal auto-destruction* had provided in the species carrying the early one-shot brain had, in effect, been rendered *useless* by the evolutionary emergence of *sleep*. So let us go back, now, to Chapter I where it is talking about hereditary attributes that have been rendered *useless*:

> When some existing hereditary attribute of a species ceases to have *survival value* ... continuing mutations and natural selection lead eventually to a *new variety* of the species whose members do *not* carry the excessive survival burden of a *then* useless attribute, as happened with what became legless sea-living mammals. If, as is generally the case, there remains a recognizable version of the attribute not *then* performing the function it did, before becoming *useless*, this is then said to be a "vestigial" attribute, and is called a "vestige."

Now, what this is implying is that, *after* the emergence of *sleep*, which rendered *useless* the *survival value* of *hereditary neuronal auto-*

*destruction*, "continuing mutations and natural selection [would have] led ... to a *new variety* of the species whose members do *not* carry the excessive survival burden of a *then* useless attribute [hereditary neuronal auto-destruction and neurodegenerative disease] ...."

This is no trivial implication, for it is saying that, *after* the evolutionary emergence of sleep, neurodegenerative disease should have become a mere *vestigial* attribute, no longer performing the function it did *before* becoming *useless*. It is saying that, in *all the species* in which a brain sleeps, *there should be no evidence of* **active** *neurodegenerative disease*!

And this brings us to a huge paradox, because we know that, at least in humans, whose brains certainly sleep, there is overwhelming evidence of active neurodegenerative disease. Then adding to the paradox is the fact that, because researchers have not found a natural "animal model" of neurodegenerative disease—an animal, *other than a human*, that can be used for research on neurodegenerative disease—it seems reasonable to conclude that the disease affects members of the human species *only*.

So there is a very important question that needs answering: **Why**—since, as evolution would have it, following the emergence of sleep, neurodegenerative disease would have become "vestigial," and there should therefore be no animals who *sleep*, in which it is *active*—**is neurodegenerative disease active in humans at all, indeed, in humans** *only*?

The beginning part of the answer is simple enough: there must have been *something* in the evolutionary development of humans *in particular* that rendered *vestigial* neurodegenerative disease *active* **again**. But this still leaves the most difficult unanswered part: **What**, *exactly*, could there have been in the evolutionary development in **humans** *particularly*—the "something"—that rendered *vestigial* neurodegenerative disease **active again** in them? To which the answer is that whatever in "the evolutionary development of humans that rendered *vestigial* neurodegenerative disease *active* again in them" must have been some development that made a change in the *nature* of the *configuration store*—in the *nature* of the human *homunculus*, as we'll now see.

# THE "INSTRUMENT" IN THE HUMAN HOMUNCULUS

It was a lovely concert, on a wintry Sunday afternoon. And now the cellist was playing, on his three-hundred year old instrument, the second of the six suites for un-accompanied cello by J. S. Bach. As he played, he'd occasionally *close his eyes*. During many of the lower-toned passages, because of the standard height of the cello which reached above his shoulder as he sat, his left hand, busy fingering strings, was too high *for his eyes to see*. It was as if the cello had become just another part of him, like an extension of his arms, hands and fingers, with no need for *eyes* to guide their placing—just as if he were touching the fingers of his left hand with those of his right, or merely touching his nose, *with his eyes closed*.

A bit envious of the cellist, I reminded myself that, after all, as I type what you're reading, I might as well be doing it *with my eyes closed*, because I'm *not* looking at where my fingers are on the keys of my keyboard—just as he wasn't looking at where his fingers were on the strings of his cello. So, we're equal—kind of, without the Bach part.

But there's a lot more to that than just Bach and envy. For, what we can see is that, whether it's a cello's strings, or keys on a computer keyboard, the fact that the player *can* **play** them **with eyes closed** is saying that, remarkable as it might seem, there must be formed, inside a human **brain**—after lots of practice, mark you—a "picture" of how fingers connect to things like cello strings and computer keyboards. Without this, there just simply couldn't be all the cello-playing and keyboard-typing...**with eyes closed**. What this is saying, then, is that, in those peculiarly **human** activities instilled, on one of evolution's aimless passages, as the *hereditary* **human** capacity to *make and use* **tools**, the continued use of these tools—"practicing" with them—can eventually lead to a "picture" of them, in our **brain**. And the "picture" must be showing them as somehow "attached" to the *earlier* "picture" in the brain that Wilder Penfield sketched and called the "homunculus," and I've been calling the "configuration store."

So, now we come back to the question: **What**, *exactly*, could there have been in the evolutionary development in **humans** *particularly*— the "something"—that rendered *vestigial* neurodegenerative disease **active again** in them?

And the answer begins as follows. As suggested by the ability of humans to execute behavior, *with their eyes closed*, in which *instruments are an integral part*, there must have been an evolutionary development that led to the motor homunculus—the configuration store—acquiring a completely *new part* that was added to its *old pre-instrument part*. And we can see how this must have happened.

The *old part* of the configuration store was what it had been when *sleep* had sufficed to allow erasure of the errors accumulating in it, thereby rendering neurodegenerative disease vestigial. This *old part* consisted of an "image" of the configuration of the actual "motor extremities" of the *body* of the host—extremities whose behavior can be controlled *with the host's eyes closed*.

The *new part* of the configuration store includes "images" of *instruments* **merged** with the *old* motor extremities in a way that allows the behavior of the extremities, ***combined with the instruments***, to be controlled *with the host's eyes closed*. This *new part* of the configuration store is what we see at work when, *with eyes closed*, the cellist coaxes the music of Bach out of his instrument, now sketched into his brain's running picture of arms, hands and fingers. This tells us the following:

> By the human host repeating—"practicing"—some *co-behavior* with an instrument, the brain carrying the configuration store can come to treat the "old" motor extremities in the homunculus "as if" they now *included* the instrument.

And now we come to how neurodegenerative disease remained in-actively vestigial, only **until** *humans* with their early tools and "instruments" came on the scene.

We begin by recalling that, when the *old* configuration store, carrying just the host's motor extremities, developed errors due to a disorder in sleep, the behavior would be limited to what just the host's *actual motor extremities* could bring on—albeit with a degree of decreased precision and control. This was followed by some later period of normal *sleep* clearing up the errors, and behavior restored to normal precision and control. *Disorders in sleep* did **not** open the door to the ***active*** neuronal auto-destruction of neurodegenerative disease.

However, with the *new addition* of **tools**—**instruments**—to the "human" host's *configuration store*, errors in the configuration store due to disorders in sleep could **now** lead to a brain directing its host's motor parts to execute behavior that might include movements that could only be executed as *co-behavior with a tool*—when there could, in general, *be no tool at all with which to execute such behavior.*

**The new "human" brain, with its instrument-augmented configuration store, has now become an entirely *new* and, effectively, "rogue" brain.**

Natural selection would not have given an emerging human species a pass to survival with such confusion in behavior, indeed, such inherent impossibility in a "rogue"- brain's directing of behavior, resulting from *disorders in sleep.*

For there to have emerged out of this situation, which natural selection would have failed to let pass, there must have been *one more* evolutionary development which left humans with a new *hereditary* endowment, such that:

> *Disorders in sleep*—sleep's quasi-absences—would lead to an *instrument-augmented* configuration store in which *errors that accumulate, due to persistent disorders in sleep,* **now** lead to *arresting the activity of such a new "rogue" brain,* by *re-activating* the *vestigial* and dormant *neuronal auto-destruction* of neurodegenerative disease.

(If a "rogue" seal wished to return to a new life on land, it would need to re-negotiate its evolutionary contracts, and have its vestigial legs re-activated, to keep natural selection at bay.)

In this way, the *errors* that accumulate, in the new, *instrument-augmented* configuration store, due to sleep-disorders in *humans,* are *once again* being treated *in the same way* as the *errors* in the configuration store had been treated—with neuronal auto-destruction—in the *very earliest* case of a *primordial* brain, which had *preceded sleep,* and never known it.

As an uncaring evolution would have it, the growing "benefit" that was brought to humans by the making of **tools,** and **co-behavior** with them, would carry with it the permanent "cost" of the *re-activation* of what had been an in-active propensity for the neuronal auto-destruction of neurodegenerative disease, if disorders in sleep should let it be.

Reminiscent of its earliest evolutionary emergence, the neuronal auto-destruction of neurodegenerative disease, now *revived* in *humans*, had become, for a ***second*** time in evolution's passage, an agent of *survival* of a species—the new, "instrumented" and increasingly "weaponized" *human* species. So, the evolutionary re-appearance of neurodegenerative disease in humans carries with it, the appearance, at the *surface*, of the *same paradox* as did its earliest primordial appearance—survivalist destruction. But, *below* the surface, this paradox, as must every paradox, finds resolution somewhere among the ceaseless efforts to "explain"—efforts of the same *surviving* human species. Evolution is too cramped for paradoxes; only the unbounded "as ifs" of the human brain has surplus space available for rent to paradoxes—until *explanation* forecloses them.

## THE "WHEN" OF IT

When did all this humanizing take place, with sleep-deprived neurons being restored to their hereditary ability to destroy themselves? The fossil record suggests this must have started about a million years back. It wended its slow evolutionary way until, some 200,000 years ago, with simple tools and instruments, and (as we'll see later) stories to match, a creature emerged that was pretty much what *we* are today—body, brain, tools and all.

Then these creatures, "as if" they saw a "future," found their Africa too small. So, leaving home, they spread as they walked, and walked, *changing* as they spread. Changing in ways that left them, at once, alike but different—different in where they lived; different in how, one day, the demographer would classify them; different in "if," then "how," neurodegenerative disease would appear in them, depending even on gender.

But it remains to wonder what behavior must be like in all those creatures with brains that *sleep*, while carrying a neurodegenerative disease only as quiet *vestige*—all those countless millions of creatures, *other than the instrumented human*, whatever their species, which host a brain that *sleeps*. What happens in them when a hereditary *disorder in sleep* occurs, with no ensuing neuronal auto-destruction, no neurodegenerative disease? What one can expect to find is a creature that displays a mildly stressful behavioral disorder resulting from a configuration store carrying small errors, until its first normal sleep erases them, restoring order. This is repeated as one sleep follows another. The creature has a brain that sleeps, better some times than others, with behavior that is ordered, some times more so than others.

And this defeats the researcher looking for the less-than-human "animal" model of neurodegenerative disease. For, if that exists at all, it will be rare indeed, hard to find, among the countless millions whose brains must sleep.

## PRECURSORS OF NEURODEGENERATIVE DISEASE

We can now see that, *in the human case*, there should be periods, between the early appearance of sleep disorders and the onset of *active* brain auto-destruction, when symptoms appear which resemble those occurring in the earlier stages of some genuine neurodegenerative diseases. These symptoms would be evidence of the growing confusion occurring in brain *motor-control* (for instance) resulting from errors in the brain's configuration store that continue un-corrected, due to ongoing disorders in sleep.

Given the origin of such symptoms, it would be reasonable to say that they are expressions of "precursors" of neurodegenerative disease "caused" by disorders in sleep, unlike genuine active neurodegenerative disease which, as explained earlier, could only reasonably be viewed as *allowed* by disorders in sleep, not *caused* by them.

One would expect these precursors to *not (yet) be degenerative*. Indeed, it would seem reasonable that these precursors might continue for years, before the underlying errors in the configuration-store become sufficiently extensive to give way to active, destructive neurodegenerative disease. There is increasing evidence that such sleep-disorder precursors exist. Dr. Postuma of McGill University has been active in this area.

## NEURODEGENERATIVE DISEASE: HEREDITARY OR NOT

As explained earlier, *sleep disorders* can be either *hereditary* or *accidental*. But we have just seen that sleep disorders are the un-guarded portal beyond which the destructiveness of neurodegenerative disease can begin. So there must be cases in which neurodegenerative disease takes hold *not* because of a *hereditary* sleep disorder, but because of a purely "accidental" one.

And this allows us to see that there must be *both* hereditary and non-hereditary neurodegenerative disease. What they will share is the unavoidably *hereditary* nature of the actual *destructiveness* of the disease; what they will *not* share is the familial ramifications that the hereditary version necessarily entails.

This now leads to an important aspect of neurodegenerative disease. For if, as just explained, there always has been an evolutionary connection between **disorders in sleep** and the eventual development of a **neurodegenerative disease**, then it must also be that those *sleep disorders* which are *themselves hereditary* will lead to *hereditary* development of neurodegenerative disease. Which leads to the following question: If neurodegenerative disease is hereditary, is this only because the sleep disorder that unleashes it is itself hereditary?

The answer would seem to be that, whereas the brain's *auto-destructiveness* in neurodegenerative disease is hereditary, the *occurrence* of the disease derives whatever hereditary nature it has from the hereditary nature of some *sleep disorders*. In effect, *hereditary* neurodegenerative disease is the outcome of the cascading of *two* aspects of the victim's genetic endowment: one expressed in the hereditary nature of the *sleep disorder* that allows the disease to take hold, and the other in the hereditary nature of the *brain-destructiveness* of the disease itself, which goes back to the very earliest evolutionary emergence of a brain.

## ENTER THE CAREGIVER

Let us go back to Chapter III, where I was introducing what must have happened when the earliest brain disabled itself. There is a passage that goes like this: "…the intensifying *stress* in the host, due to the increasingly confused direction of its parts, brings on a *new brain-condition*…in which the now offending brain is *disabled*, by having the working of its neurons *destroyed*. **The result of this is that the host then continues to live without the benefit of a working brain, much as the host's *multi-celled evolutionary antecedents* had done, without having neurons at all.**" (I've added the boldface here.)

As we can now see, that is exactly the condition brought on, as neurodegenerative disease continues to take hold—the victim will, in effect, be eventually set back some eons of evolutionary time. For, as the primary expression of the disease, the brain will eventually have destroyed the working of its *own* neuronal elements. And, as it says in the boldfaced passage: "The result of this is that the host then continues to live without the benefit of a working brain, **much as the host's *multi-celled evolutionary antecedents* had done**, without having neurons at all."

But humans cannot survive, living *alone*, as their *multi-celled evolutionary antecedents* had done, and still do, without having working neurons at all. For what, in our situation, is significantly different is that, while those *multi-celled evolutionary antecedents* can continue to feed and get rid of their wastes all by themselves, we—the relatively complex feeders and waste eliminators we've become—are unable to do so in a brainless condition. And, since this is the state to which neurodegenerative disease will eventually carry them, humans will need the growing attention and support of the *caregiver*.

Then, in a marvelously human display, the human victim and caregiver, together now, *outdo* their earlier brainless multi-celled ancestors by joining up into a resilient kind of completely new, *extended*, multi-celled creature, with what is now approaching a single shared brain between them, directing and sleeping, in exhausting defiance of where an ancient evolutionary disorder would take this newly unified "them."

But there's more to it than that.

# CHAPTER VI
# WHEN
# NEURODEGENERATIVE
# DISEASE TAKES HOLD

## WHERE THE ANSWER LIES

To answer the question of what happens when neurodegenerative disease takes hold, we need to begin by going back to Chapter V, where it says:

> … disastrously for the host, when there are *sleep disorders*, and auto-disabling has **begun** in this new kind of brain that normally sleeps, then the **more** *it is driven* to auto-disable itself, **still more** is it driven to auto-disable itself—auto-disable the **neural** components that make it up—on and on, **degeneratively**. So, when such destructive brain auto-disabling commences, we are witnessing the onset of *a "neuro-degenerative disease,"* as we now call it—one which had been **preceded by disorders in sleep**, and has its primary focus in the **degenerative auto-destruction of the neuronal components that make up a brain**.

## THE GENERAL ANSWER

The general answer comes directly out of the last sentence:

> ... when such destructive brain auto-disabling commences, we are witnessing the onset of *a "neuro-degenerative disease,"* as we now call it—one which had been **preceded by disorders in sleep,** and has its primary focus in the **degenerative auto-destruction of the neuronal components that make up a brain.**

So the general answer, stark as it is, is as follows:

> Neurodegenerative disease involves the degenerative auto-destruction of the neuronal components that make up a brain. Therefore, as the degeneration progresses, **all the functions of a brain, whether "*sensorial,*" or "*mental,*" or "*directorial,*" could eventually fail.** "Sensorial" functions are those associated with the brain's involvement in sensation, as in hearing, taste, touch, smell, and sight. "Mental" functions are those associated with brain-originated manifestations in the brain's host such as memory, imagination, "standing for," "as if," and so on. "Directorial" functions are those associated with the brain's direction of motor parts of its host: legs, hands, feet, tongue, for example.

This is the "General Answer," as I'll refer to it in future. But let us look, now, at some of the details to which it leads.

## DETAILS TO WHICH THE GENERAL ANSWER LEADS

The General Answer suggests that there would be "three different kinds" among the various neurodegenerative diseases—as their origins would be "*sensorial,*" or "*mental,*" or "*directorial.*" But, because of the interconnectedness of the areas of the brain, it is more likely that, as the actual clinical evidence suggests, there would be overlap and merging of the various "kinds" of the diseases, beginning even at their earliest emergence, and becoming more merged as the degeneration advances.

However, the evolutionary origin of neurodegenerative diseases allows at least a surmise as to some "natural" order in which they might *tend* to emerge, based on the evolutionary order in which the *neuron* must have acquired its behavioral attributes (which I'll get back to later). This would be a "natural" order in which the **sensorial** aspects of the diseases—associated with the brain's involvement in sensation, as in hearing, taste, touch, smell, and sight—would appear *first*. These would be followed by the **mental** aspects—associated with brain-originated manifestations in the brain's host such as memory, imagination, "standing for," "as if," and so on—which would appear **second**. The last aspects to appear would be associated with the **directorial** function—associated with the brain's direction of parts of its host: legs, hands, feet, tongue, for example—which would appear **third**. But before discussing this natural order, it will be useful to look briefly at "pain."

## PAIN AND EMPATHY

Pain is a "feeling" that signifies the existence of some abnormal condition in a creature, often of damage that has occurred. And, for the present discussion of what happens when neurodegenerative disease takes hold, it is useful to recall that this "*feeling* of pain" is located in the *brain—not* at the place where the abnormal condition has occurred—and is mediated by *neurons in the brain*. Consequently, destruction of the brain's neurons will destroy the "*feeling* of pain."

The evolutionary emergence of the "*feeling* of pain" must have had crucial significance for survival, for it tends to prevent the repeating of behaviors that cause abnormal conditions, as evidenced in "once bitten, twice shy."

All the foregoing applies to the "*feeling* of pain" as it occurs in the creature in which the abnormal condition has *actually* occurred. But it is possible, certainly among humans, for one to "share" the "*feeling* of pain" occurring in another. This is generally taken as a manifestation of "empathy." And empathy, also, must have had crucial significance for survival, since it tends to prevent repeating of behaviors that caused an abnormal condition in *some other* similar creature.

One might therefore wonder what, in a brain, would lead to, for instance, *my* feeling of *your* pain, *as if* it were my own. Well, the answer would seem to be right there in the *as if*—there, where as it says above, in the General Answer, that the brain's "mental" functions are associated with brain-originated manifestations in the brain's host such as memory, imagination, "standing for," **"as if,"** and so on. From which we can draw the following important conclusion:

If neurodegenerative disease should destroy the neurons supporting a brain's *mental* functions, then both the "*feeling* of one's own pain" and the "*feeling* of another's pain"—empathy—will fail. And, those *constraints on behavior*, which might have come from either the "*feeling* of one's own pain" or the "*feeling* of another's pain," will vanish with the destruction of the neurons supporting a brain's *mental* functions, as described in the General Answer. (I return to this subject later, in Chapter VIII, in the context of narration.)

So, let us now return to the "natural" order in neurodegenerative diseases, raised by the earlier surmise, to see what that might entail.

## WHAT A "NATURAL" ORDER IN NEURODEGENERATIVE DISEASES ENTAILS

That there could be a "natural" order of emergence of neurodegenerative diseases is of particular interest because, appearing as the earliest in such an order, the **sensorial** ones would tend to be seen as just "ordinary," "normal" losses of sensory capacities, especially in older people—just complaining, perhaps, of a little trouble sleeping; not quite hearing what grand-children are saying; visiting the ophthalmologist who, having looked eye to eye, scribbles "glaucoma"; coffee that just doesn't taste so good anymore! Much more seriously, though, because of their evident "ordinariness," they would not be seen as the *tell-tale signs of the very beginning of neurodegenerative disease* they could be.

But if, in the "natural" order, these were then followed by the **mental** aspects of the diseases, these would not seem at all ordinary or normal. Because *loss of memory*, or what a word "stands for," could not be seen as normal, with such a loss entailing an evident reduction in the capacity of the victim to lead a normal life. This would be part of the family of diseases, "dementias," which includes "Alzheimer's" disease. Furthermore, because of the loss of the ability to indulge in "as if," the dysfunctions would include a loss of the ability to play, to enjoy "pretending" "as if" something is not what it really is—the joke, the riddle is no longer "funny." But more sobering is the loss of being able to feel "as if" one were another person, thereby, as explained earlier, entailing the loss of the capacity to feel "as if" someone else's *pain* is one's own—the loss of empathy. This then ushers in the evident danger that the victim of this phase of the disease, still physically strong—and even once gentle—can now pose to *others*.

With the aspects of the diseases to appear *last* being those associated with loss of the brain's **directorial** functions, these would become readily apparent to just simple observation as loss of control in the victim's larger and smaller motor parts—the inability to hold a hand steadily in a single place; to touch a nose with eyes closed; to move a leg, so the wheelchair is now a partner in life. Here are located the family of diseases of which Parkinson's is the most common. And this first-to-last order of the diseases tells us why, when still afflicted with Alzheimer's, but not yet with the paralyzing disabilities of Parkinson's, the victim—still fully mobile, but having lost *memory*—can become unable to find the way home … from just down the road.

But there would be more subtle aspects of this last phase of the disease. For there exist what, clearly, are basically *instinctive* functions, not relying *primarily* on brain direction, but which, certainly in humans, have become subject to at least *limited* brain direction. Some of these are breathing, swallowing, urination, and defecation. So it could be expected that, in this last phase of the diseases, there would be loss of control of breathing that could become apparent in loss of speech, for instance; loss of control of swallowing that would pose problems in feeding; loss of control of urination and defecation that would pose problems of incontinence.

From all this, we can see that, especially in its third and final phase, neurodegenerative disease could be expected to display what would have become the *overlay* of its later, onto its earlier phases. So the victim's condition could then involve deafness or blindness, overlaid with the memory-loss of Alzheimer's, overlaid with the paralysis and incontinence of Parkinson's, in what will have become the ultimate test of the affection, patience and sheer physical strength of the caregiver.

For, as explained earlier, with assistance, the loss of the functions of a brain will not necessarily lead to death in a human. All the bodily functions that had had earlier evolutionary emergence can continue to survive, if provided with feeding and removal of waste. So, in the tireless skilled hands of the caregiver, neurodegenerative diseases are not *themselves* generally fatal. However, the chronically inactive condition imposed by the complex latest phase of the disease leaves the victim a relatively easy target for *infection*, which can finally bring death.

## Back To the Neuron and Its Attributes

The surmise in an earlier section was that the way the neuron evolved could now determine a "natural" order in which neurodegenerative diseases develop. And since, as just shown, such a "natural" order helps to explain why neurodegenerative diseases progress the way they do, it would be useful to see if, indeed, the evolution of the neuron might have had anything to do with a "natural" order. Let us, therefore, go back to Chapter II, in which we find, with additions that I've placed between a [and a]:

> So, there's something *essentially new and very different* about this kind of transfer between these evolutionarily later-occurring kinds of stretched-out cells [identified a few paragraphs later as "neurons"]. For, although it is still *molecules* that are transferred between the first and second cells, *what* is being *conveyed* by *this* transfer is not molecules functioning *as just the molecules they are*, but, more importantly, it is molecules functioning as something *different* from what they are. They are no longer molecules involved in just chemistry: they are molecules involved in something *more*. For what is now happening is the passage between cells of a molecule that is not *itself* the "increase in pressure," [that the neuron had "sensed"] but a molecule that is "standing for" it—a molecule that is not the "happening itself" [that the neuron had, in the first instance, "sensed"] but a molecule that *stands for* the happening itself [with the "stands for" coming, necessarily, only *after* what had been "sensed"].

Already we can see here that, in the neuron—which, as evolution would have it, neurodegenerative disease eventually destroys—"sensing" came *before* "standing for." Consequently, the order in which a brain's disabling would proceed would have started with the disabling of "sensing," before it could have come on the disabling of "standing for"—which is as the earlier surmise would have it.

In the same way, we could go back to an earlier chapter to show that *direction* of a host's parts by a brain must have emerged only *after* the neuron had stretched out into various parts of its host. But this seems unnecessary, for, even without this, we can see that there does, indeed, appear to have been a "natural" order in which the phases of neurodegenerative disease would proceed: *sensorial, mental, directorial*—an order that reflects what must have been the very early evolutionary history of the neuron.

And here I must confess *my own* surprise that looking at such an obscure little piece of neuronal evolutionary history could even *hint* at why the victim of Alzheimer's, having, in a *natural* order, lost memory *before* being physically disabled by Parkinson's, can wander and become lost; why the victim of Alzheimer's, having, in a *natural* order, lost the capacity to "as if" but *before* becoming physically disabled by Parkinson's, can lose the constraining effect of empathy, and become a present danger to others; why, in the latest phase of neurodegenerative disease, the victim of the physical un-control of Parkinson's is the near-silent, un-recalling wheelchair-bound victim of what is really the *superposition of the phases* of the disease, unforgiving phases as they are, that came in their *natural* order.

But all these are expressions, in neurodegenerative disease's unforgiving ways, of sleeping brains hosted in *humans alone*. So, could there be yet *other* expressions of a sleeping brain, hosted in humans alone, which, if we looked closely enough, would *also* show themselves as the ways of neurodegenerative disease—ways even just as unforgiving? Let us look.

# CHAPTER VII
# ADDICTION AS
# NEURODEGENERATIVE
# DISEASE

## WHAT "ADDICTION" MEANS

There are numerous versions of what "addiction" should be taken to mean. From among these, I will be taking the word "addiction" to mean the following:

> "Addiction" is a condition that afflicts humans. In the presence of this condition, the afflicted, the "addict," engages in behavior that recurs, despite consequences deemed harmful, by onlookers, to the health, mental state, or social life of the addict.

## ABOUT TENABLE EXPLANATION

Addiction can have devastating effects on the addict and on those affected by the addict's behavior. As a result, addiction has been the focus of a great deal of attention: research, clinical, social, curative, legislative, legal, punitive, and more. Growing out of this activity, there have been numerous "explanations." But their conflicting variety testifies to their inadequacy in providing *tenable* explanations of either what causes addiction, or how it might be cured, if, indeed, this were possible. And since I am here going to be presenting an explanation of addiction, it seemed desirable to review what would constitute one that might be *tenable*, much as given for tenable explanation in Chapter I.

I begin with excluding all those explanations that explain addiction *only*. The reason for this is that any "single," "separate" phenomenon can be explained in many, perhaps unlimitedly many ways. So the proper test of an explanation is not simply: Can it explain some phenomenon? Instead, given that an explanation explains some phenomenon, the real test of whether it might be tenable is: Can it explain *something else*—especially something else situated just "close by"? And, indeed, the *more* other *something elses* to which the *same* explanation can be extended, the *more tenable* the explanation would be.

By way of example: That the Sun is a glowing chariot driven across the sky of a stationary Earth is a perfectly good explanation of what the Sun can be seen as doing—it needn't even be the same driver and chariot, day to day. But the explanation is untenable, and what is *sufficient* to make it so is its inability to explain Earth's *seasons*: a "something else" the Sun is also distinctly involved in, as we can clearly see.

So, having excluded explanations that explain just addiction, and *nothing else*, there is the further requirement that the explanation should lead to possible confirmation or denial based on the extent to which it leads to what is known about addiction—should be "falsifiable"—or, even better, leads to what can be *predicted* about addiction that might yet be brought to light.

In summary, then, there is the requirement, for an explanation to be tenable, that its explanatory power should not be confined to just the immediacy of what it sets out to explain, and that the explanation should be "falsifiable." With that, let us see, now, if the explanation of addiction as another of the many manifestations of neurodegenerative disease seems tenable.

## SOME PRELIMINARIES

In order to simplify the later explanation of addiction itself, I'm going to discuss a number of subjects that, as you will see, are very much connected to addiction. They are the following: "the future," "satisfaction," "need," "compulsion" and "consequence." This will be done in the context of neurodegenerative disease, as it has been explained so far.

These preliminaries might seem a bit tedious, but read on. They lead to interesting places.

## "THE FUTURE" AND THE "AS IF" OF NEURONS

To see what the relationship might be between "the future" and the "as if" of neurons, it will be convenient to start with the "present." The present is characterized by what is going on now, by what can be *sensed directly*—seen, heard, tasted, touched, smelled—and moved, via the "as if's" of neurons, into a brain. But "the future" is fundamentally *different* in that it *can never be sensed directly*—it isn't "there yet" to be sensed directly—and so "the future" cannot be carried, even by the "as if's" of neurons, into a brain, as the present can. Consequently, to the extent there is "the future" in a brain at all, it must have been *created,* all of it, right there, *by neurons in the brain.* From which we can conclude the following:

> When, in a brain, the neuronal auto-erasure of neurodegenerative disease has set in, "the future" will be *erased* with the neurons that created and supported it.

We can summarize this in the following sequence:

> neurodegenerative disease → neuronal auto-erasure → erasure of "the future."

## SATISFACTION

Oh, that piece of chocolate tasted so good!

But although the "good" taste came right then and there, "satisfaction," if it came *at all,* came only *after* the tasting. And that's the way satisfaction has always been: coming *after* the tasting, or the hearing, or the seeing, or the touching.

Indeed, there's no way to *sense* "satisfaction" *directly* at all. So, there's nothing of "satisfaction" to be *carried* by neurons *into* a brain. Consequently, whatever there is of "satisfaction" *in* a brain must have been made right there, by the neurons of the brain themselves.

From which we can conclude the following:

When, in a brain, the neuronal auto-erasure of neurodegenerative disease has set in, "satisfaction" will be erased, and the brain's host will become "un-satisfiable."

We can summarize this in the following sequence:

neurodegenerative disease → neuronal auto-erasure → un-satisfiable.

Un-satisfiable? I've used this strange word because, at the end of the sequence, there will be no "satisfaction" *possible*—the person afflicted with neurodegenerative disease is then *never* "satisfied," is then *beyond* "satisfaction," is "un-satisfiable."

But if this is so, what will behavior look like when the person afflicted with neurodegenerative disease is "un-satisfiable"?

## RECURRING BEHAVIOR, NOT "TRYING": "COMPULSION"

What, indeed, will behavior look like when, as concluded above: "the person afflicted with neurodegenerative disease is ... 'un-satisfiable'"?

To begin an answer, we can notice that "satisfaction" allows the host of any *normal* human behavior to cease that form of behavior, and go on to something else. Indeed, "satisfaction," that announcer of "enough," is essential to the control of an aspect of "normal" behavior—when to go on to something else, or just when to cease.

What we are led to conclude, therefore, is the following:

When the afflicted with neurodegenerative disease is *beyond* "satisfaction," is "un-satisfiable"—as just shown it can become—behavior arrives at a condition in which, being **un**-able to arrive at a normal *stable* state of *satisfaction*, it *recurs* (the physicist would say it *oscillates*), as a now *diseased* brain searches for what would be "normal" "satisfaction," where none is to be had—stabilizing "satisfaction" having been erased *by the disease itself.*

This "recurring" behavior is what appears *to the onlooker* as "compulsive" behavior, or simply as "compulsion." But what, for the *onlooker*, is "compulsion," perhaps even an act of "will" is, for the *afflicted*, simply behavior that "keeps on repeating itself." Significantly, the afflicted *is not "trying" to perform the "recurring" behavior*. Instead, the afflicted is executing "recurring" behavior which, having had no way of arriving at satisfaction, keeps on recurring, until the host's "un-satisfiable" brain carries the host to some new condition, which arrests it.

We can summarize this in the following sequence:

> neurodegenerative disease → neuronal auto-erasure → un-satisfiable → not "trying," recurring (compulsive) behavior.

But we can go further by making use of something that was said earlier, in Chapter V:

> It becomes evident, from all this [preceding discussion of disorders in sleep], that there is, and always has been, *an evolutionary connection* between **disorders in sleep**—amounting to its quasi-absence—*and the development of a destructive* **neurodegenerative disease.** ... [T]hey become the un-guarded portal beyond which the destructiveness of neurodegenerative disease can begin.

And we can go further because this last quote, when combined with what was just concluded about "un-satisfiable," gives us another of those sequences:

> disorders in sleep → neurodegenerative disease → neuronal auto-erasure → un-satisfiable → not "trying," recurring (compulsive) behavior.

*Keith C. M. Glegg*

## "NEED": AT THE OTHER LIMIT FROM "SATISFACTION"

To get at the next item in the preliminaries mentioned earlier, let's go back further than just that piece of chocolate, to "eating," more generally. What becomes evident is that, a species whose members did not carry, as part of their genetic endowment, the *upper* limits on eating, set by *satisfaction*, would, long ago, have had extinction for its *greedy* fate.

But if evolution must have set a limit on eating "too much," it must also have set one on eating "too little," since a species whose members did not carry, as part of their genetic endowment, the *lower* limits on eating would, long ago, have had extinction for its *starving* fate. This lower limit is set by what we generally refer to as "need."

All the previous conclusions about "satisfaction," reached earlier, and summarized in the sequence immediately above, apply to "need," so I needn't repeat them here. (As a little exercise, you might want to run through them yourself.) What is lacking is a word, equivalent to "un-satisfiable," which will mean that, where there should normally be a *lower* limit, behavior is no longer limited by "need." For want of anything better, I've settled on "un-needing." This then, with the previous sequence, gives the following:

disorders in sleep → neurodegenerative disease → neuronal auto-erasure → "un-satisfiable"↑ *or* "un-needing"↓ → not "trying," recurring (compulsive) behavior.

I've added the symbols ↑ and ↓ as reminders that "satisfaction" acts as a limiter at the *upper* end of behavior—eating, for instance—and "need" at the *lower* end.

So, "normal" behavior, like eating, has a *range,* in which behavior passes back and forth, with *satisfaction* setting the upper and *need* setting the lower limit—a range for some narrower than for others, for some, wider.

## Consequence

This is the last item in the preliminaries. To begin, we notice there is an *act* in the *present*, then, in "the future," there is its "consequence." There is nothing *in the act itself* that *is* its consequence—consequence, if there's to be any, can emerge in "the future," only. Therefore, if, in a host's brain, the neuronal auto-destruction of neurodegenerative disease should conspire to erase the "the future," *consequence* would be erased with it, and the host's acts would then *lack* "consequence." Indeed, acts in such a "futureless" host, *whatever* the acts, would, for the *host*, be lacking consequence, would be "in-consequential."

We can therefore go back to the sequence which summarizes what had just been concluded about "need":

> disorders in sleep $\rightarrow$ neurodegenerative disease $\rightarrow$ neuronal auto-erasure $\rightarrow$ "un-satisfiable"↑ *or* "un-needing"↓ $\rightarrow$ not "trying," recurring (compulsive) behavior;

then simply add to it what has just been concluded about "consequence," to yield the following final sequence:

> disorders in sleep $\rightarrow$ neurodegenerative disease $\rightarrow$ neuronal auto-erasure $\rightarrow$ "un-satisfiable"↑ *or* "un-needing"↓ $\rightarrow$ not "trying," recurring (compulsive) behavior $\rightarrow$ "in-consequential" behavior.

This ends the preliminaries. Let us now look at what they mean for addiction.

## How to Make an Addict

"Addiction" is a condition that afflicts humans. In the presence of this condition, the afflicted, the "addict," engages in behavior that recurs, despite consequences deemed harmful, by onlookers, to the health, mental state, or social life of the addict.

So, how could we make an addict? Well, we're going to be guided by the very last sequence, above.

We could start with someone who says he's been having trouble with getting a good night's sleep—just can't seem to "get to sleep." But he still enjoys his glass of wine with supper. That's it, just one glass of wine with supper, because that *satisfies* him—just one glass. But if we had a way of making just one glass satisfy him *no longer*, he might get to two glasses, or even a bottle, still *without* him being satisfied. So, now he tends to be drunk all evening, and even has another half-bottle in the morning, still not satisfied—we've made him *un-satisfiable* by the drinking of wine. Even better, we now have him not "trying" to drink himself drunk—according to *him*, it just "recurs." All of which, so far as we, the *onlookers* can tell, causes lots of troubles, both at home and at work—social troubles.

Well, the increasing troubles might have caused him to go back to just a glass with supper, but, since we're still trying to make an addict of him, we consult the sequence, and arrange to have him *not care* about the *consequences* of being drunk much of the time. And we've succeeded pretty well, because now he's drinking wine even more, *un-satisfiable* as he's become, drinking just *recurring*, and all the increasing social troubles are now, for *him* anyway, *in-consequential*.

So, what did we do to so transform this one-glass wine drinker into an addict, an alcoholic—a "winebibber"? Well, maybe *we* couldn't *actually* have done it. But to see what must have happened, that we've been taking too much credit for, let's go back to our guiding sequence:

> disorders in sleep → neurodegenerative disease → neuronal auto-erasure → "un-satisfiable"↑ *or* "un-needing"↓ → not "trying," recurring (compulsive) behavior → "in-consequential" behavior.

What this is telling us is that, to be "making an addict," what we have to do is begin with someone afflicted with neurodegenerative disease in one of its many early forms, someone with the right (wrong) heredity, or just the right accident—troubles of some kind—that would make for disorders in sleep, but better if both; then watch for the *un-satisfiable,* and then ... just wait. "Compulsive" behavior, with its festering grief, as the *onlooker* would have it, but all without "trying," all "in-consequential," as the *addict* would have it, *will be along*—not in *every* case, mind you, but often enough to suggest that the foregoing sequence might just be tenable.

But this is only one kind of addict, what other kinds might there be? To answer the question at all fully, we'll need to have a brief look at "learned" behavior.

## "LEARNED" BEHAVIOR AND KINDS OF ADDICTS

What is a "learned" behavior? To see what a learned behavior is, let us begin with a very new little *mammal*—pig, elephant, rat, whatever— just born, which must *suck* in order to survive. No species of mammal could survive if its members, at birth, had to be *shown* **how** to *suck*. Members of every surviving mammal species come *knowing* **how** to suck, and that includes the members of our own species—with their remarkable early liking for thumbs, as well. So we come with some kinds of behavior, like sucking, that we don't need to be *shown* **how** to perform, we just perform them.

But that certainly isn't the case with *all* our behaviors. For instance, to speak whatever language we speak, we have to hear someone else speak it. So, even if our parents speak Chinese, if what we hear is only French, we grow up speaking French. This isn't like sucking, where we suck at birth just as our parents, and their parents sucked at birth. Instead, we speak whatever we speak because that's what we "learned" to speak, unlike sucking where we didn't have to "learn" anything, we just did it.

So "learned" behavior is behavior that we "catch" from others, when and where we catch it, if, indeed we do. (*How* we come to "catch" it I leave to a Chapter VIII.) What we can be sure of, though, is that, although we don't come with the language(s) we speak, it must be that we all come with *whatever it takes to "learn"* the language(s) we speak, since, with unfortunate exceptions, we all speak some language(s) or other.

Let us now go back to sucking. And we can see that sucking, like all other normal behaviors, needs to have some way of arriving at being "satisfied." But since we come all equipped with sucking, not having to learn it, sucking must come all equipped with *satisfaction* built in, not having to learn that either. And this must be the same with all the other behaviors that we come with, not having to learn them: swallowing, digesting, breathing, urinating, and so on. All these behaviors must have their "satisfaction" built in, and it must be the same with "need."

By comparison, **none** of the behaviors that are **learned** could come with their *satisfaction* or *need* **built in.** *Their* satisfactions have to be *learned* as well—as the behaviors are learned.

And it is their **necessary learning** which places the **satisfactions and needs** attached to **learned** behaviors *within the* **neurons of a brain**.

Consequently—and this is a crucial consequence—we have the following conclusion:

> When, as the neuronal-auto erasure of neurodegenerative disease advances, "satisfaction" and "need" are erased in a brain, it is "learned" behaviors that will become the behaviors of the "un-satisfiable," at the one extreme, and the behaviors of the "un-needing" at the other. Therefore, it is "learned" behaviors, in their *unbounded* form, that will become the behaviors of the addict, which, *for the addict,* simply continue to "recur," without "trying," and without "consequence." Therefore: **For every kind of "learned" human behavior there will be a kind of addict.**

We can summarize this conclusion, which I'll call the "learned addiction conclusion," by extending the previous sequence as follows:

> disorders in sleep → neurodegenerative disease → neuronal auto-erasure → "un-satisfiable" (f) "learned behavior"↑ *or* "un-needing" (f) "learned behavior"↓ → not "trying," recurring (compulsive) behavior → "in-consequential" behavior.

You will have noticed that "learned behavior" has been added to both "un-satisfiable"↑ and "un-needing"↓ in the previous sequence by writing:

"un-satisfiable" (f) "learned behavior"↑,

and

"un-needing" (f) "learned behavior"↓.

The (f) is simply intended to indicate that it is "learned behavior" that will characterize the "un-satisfiable," and similarly for the "un-needing." (The mathematicians would say that the (f) means "is a function of," which is close, so we'll let them have their way.)

There is just one other aspect of "learned behavior" that needs mentioning here. This concerns those behaviors that, in infants, must be like sucking—not needing to be learned from others—but which, in the course of normal development, acquire *learned* components. These are behaviors such as breathing, urinating, defecating, eating, walking, and sleeping.

For instance, a new-born must be able to breathe simply by virtue of having entered a world containing free oxygen, so, once started, the infant breathes the way it sucks—"automatically." However, in the course of normal development, the infant will learn to "hold" its breath; to issue pulses of breath that become syllables on the way to speech; to not suffocate while it sticks out a mocking tongue, and so on.

So, let's go back to the "learned addiction conclusion," where it says:

**For every kind of "learned" human behavior there will be a kind of addict.**

For, given what has just been said about breathing, for instance, the range of "learned" behaviors has evidently been considerably enlarged. So we can now expect to find that the behaviors of addicts include not only those which clearly must be learned, like language, but, in addition, all those other behaviors which, although in their earliest forms must be "automatic," have acquired some *learned* aspect, as evolution and natural selection will have ordained.

It's now time to return to answering the question that, in referring to the "wine bibber," introduced this discussion of learned behaviors, by asking: "But this is only one kind of addict, what other kinds might there be?"

## KINDS OF ADDICTS

We can begin by recalling that, if the neuronal auto-destruction of neurodegenerative disease should *erase satisfaction* or *need*, as it can, then behavior might cease to display its normal upper limit, or its lower, or both. And, with that, it might also erase *consequence*, as it can—all in the one brain that evolution, in a long-ago survivalist moment, had taught to destroy itself.

So there will be *un-satisfiable* addicts that engage in *learned* behavior that *recurs*, despite *consequences* deemed harmful, by *onlookers*, to the health, mental state, or social life of the addict—sometimes behavior *above* satisfaction's normally upper limit; sometimes behavior *below* need's normally lower limit—in:

> eating, taking opiates, walking, talking, and sleeping, betting, mating, and taking, and giving, and paying attention, associating, praying, forgiving, and searching, and locking doors, sheltering, and speeding, and hurrying, verifying, washing hands, tidying, and fixing personal features, buying, selling, and making lists....

Some few of these have names: anorexia, bulimia, heroin addiction, for example. With anorexia and bulimia, both eating addictions, there is, as there always must be, the one possible addiction *below* the "normal" *needing* level, the other *above* the "normal" *satisfaction* level—eating too little, or too much; getting too meager, or vomiting while not. Then there is "associating," which yields the "loner" at the un-needing extreme, and the gregarious "mixer" at the un-satisfiable extreme. We'll be meeting the "loner" often.

But, as might now be seen, most addictions would be as individually un-nameable as they would be individually un-identifiable, being, as they are, no fewer than the un-countably many varieties of entwined components of normal human learned behavior.

## MANSIONS AND SIDEWALKS

The ways in which humans provide shelter for themselves vary widely. Indeed, even today, some humans live in climates in which shelter is so close to unnecessary, as to be almost non-existent. So we can be certain that the *building* of shelter, especially as this takes the form of *house-building*, is a *learned* behavior. But the "learned addiction conclusion" is saying:

> [I]t is "learned" behaviors, in their *unbounded* form,
> that will become the behaviors of the addict, that,
> *for the addict*, simply continue to "recur," without
> "trying," and without "consequence"; that will become
> the behaviors that appear when, as the neuronal-
> auto erasure of neurodegenerative disease advances,
> "satisfaction" and "need" are erased in a brain.

What this is telling us, then, is that house-building, for "shelter," can become the behavior of an addict. So, at the "un-satisfiable" end, it can be expressed in the "mansion" that goes well beyond the need for just shelter, all the way to the fourth, or seventh or even tenth "house" for "shelter." The behavior, *for the addict*, simply continues to "recur," without "trying," and without "consequence." It is the behavior that appears when, as the neuronal-auto erasure of neurodegenerative disease advances, "satisfaction" is erased in a brain. But what if it were "needing," not "satisfaction," that had been erased?

This is at the other extreme, and produces the addict who is "un-needing" of shelter. It is then the no-shelter of the *sidewalk*—being as far as the urban addict, this un-needing "loner," can go in avoiding shelter—that becomes the focus of behavior. For the *addict*, the no-shelter of the sidewalk simply continues to "recur," without "trying," and without "consequence." It is the behavior that appears when, as the neuronal-auto erasure of neurodegenerative disease advances, "need" is erased in a brain.

How curious, indeed, that the failing neuronal "as ifs" in a brain can exchange the nice leather brief-case, carrying those "important contracts," for just a plain paper bag carrying ... all the belongings of a "homeless," loner addict.

All this, if *satisfaction*, at the *upper* limit or, or *need* at the lower limit should be erased by neurodegenerative disease. But what if neurodegenerative disease should erase *satisfaction* and *need* **together**, leaving neither an upper nor lower limit to what had been "normal" behavior? What would behavior look like then, to the onlooker?

## TWO KINDS OF ADDICTION IN ONE

What, indeed, would behavior look like, to the onlooker, if neurodegenerative disease should erase *satisfaction* and *need* **together**, leaving neither an upper nor lower limit to once "normal" behavior?

Well, with *satisfaction* and *need* erased **together**, and behavior that *recurs,* still, to the *addict,* being *inconsequential,* there are no longer either upper or lower bounds on what recurring behavior can become. So, there can now be *two* kinds of recurring behavior, *two kinds of behavior merged into one addiction*: all the way from behavior *above* "satisfaction's" *once* normal—excessive, grandiose, "high"—to behavior *below* "need's" *once* normal—meek, "loner," depressed—always *passing through* normal, on the way from one now un-bounded extreme to the other.

Even the not-so-careful onlooker could give this kind of behavior a fitting name, moving as it does from one extreme to the other. A fitting one would be "bi-polar" behavior—two extremes, two "poles." But this isn't how the onlooker saw the behavior at first. At first the onlooker saw it as consisting of "manic" episodes, at the upper extreme, and episodes of "depression," at the lower extreme. So, for a long time, this two-state addiction was known as "manic-depressive," until that was replaced by the more descriptive "bi-polar."

For those who know even a little of this story's history, it will doubtless be surprising that the explanation of bi-polar "mental illness" should be made, as it is here, to fall within a *single* explanation of *addiction,* which *itself* falls within the *single* explanation of the evolutionary nature of *neurodegenerative disease.* Indeed, that manic depression—which has had so many unrewarding attempts at its explanation—should reveal itself as just neurodegenerative disease, expressing itself in an extreme bi-polar addiction, surprised ... even me! So, let us pursue this—interesting as it's turned out to be—a little further.

## SERIAL KILLING

At one extreme of bi-polar addiction—the "manic" extreme—there is recurring behavior that is "excessive, grandiose, high." But, now, we have that this will be *learned* behavior. So, what, among learned behaviors, could be more excessive, or grandiose than killing a human? Of course, for the now extreme *addict,* this would be without "trying," just "recurring," and "in-consequential"; but for the *onlooker*, and indeed for the *victim* consequential in the extreme, deadly, perhaps grossly full of pain, because of all the "learning" that the addict could bring to killing.

There could be just one killing, followed by a passage through normal behavior, with the addict headed for the other pole of addiction—the "depressive" pole. But if depression had spared the addict—and, as we'll see, it not always will—there is a return, through normal behavior, to the other extreme, excessive, grandiose, high, as before, with *another killing*, in what is now a repeating cycle: passage from one extreme to the other, through normal behavior, with what will appear, to the *onlooker*, like nothing more than another "serial killing." (There is the question of how the serial killer could enact killing, time after time, without being identified and restrained by onlookers. This will be answered in the following section.)

But what might behavior become when the bi-polar addict, having passed through a period of normal behavior, arrives at the other pole, the one determined by "un-needing"?

## SUICIDE AND "UN-NEEDING"

Let us consider those *learned* behaviors that *advantage* preservation of life, for these are behaviors which—whatever form evolution and natural selection still take in humans—must have become a crucial part of the behavior of members of a surviving human species. Among these learned life-preserving behaviors we find: avoidance of falling from high places; avoidance of drowning; avoidance of ingesting known or suspected poisons; avoidance of strangulation; avoidance of self injury with tools and weapons; avoidance of swallowing in-digestible objects; avoidance of harm due to enemies and predators, among others. How might these be expressed in the behavior of addicts?

At the one extreme, they will appear as behavior of the un-satisfiable. So they will be behavior that involves *excessive* enactments of all the behaviors just mentioned—excessive *avoidance* of falling from high places; of drowning; of ingesting known or suspected poisons; of strangulation; of self injury with tools and weapons; of swallowing in-digestible objects; and of harm due to enemies and predators. For the "un-satisfiable" *addict*, all this behavior involving *excessive avoidance* simply continues to "recur," without "trying," and without "consequence." However, for the *onlooker*, the behavior of the un-satisfiable addict is simply "cautious"—even "decent" and cautious. (And here resides the answer to the question of how the serial killer could enact killing, time after time, without being identified and restrained by onlookers. Because it is what the *onlooker sees* as "cautious" behavior in the *addict* which *allows* the serial killer, as deadly "loner," to enact a fatal serial—un-seen, un-suspected, and un-restrained.)

At the other extreme, the addict becomes "un-needing," and the learned behaviors that normally *advantage* preservation of life **cease** being "needed." So behavior falls *excessively below* the normal "needed"—the **un-needing** addict **will not** *avoid* either falling from high places; or drowning; or ingesting known or suspected poisons; or strangulation; or self injury with tools and weapons; or swallowing in-digestible objects; or harm due to enemies and predators. For the "un-needing" *addict*, all this behavior involving **non**-avoidance simply continues to "recur," without "trying," and without "consequence." However, for the *onlooker*, the behavior of the un-needing addict becomes excessively "**in**-cautious."

But we have also seen that the un-needing addict can display typical behavior *below* "need's" *once* normal: meek, "loner," depressed. So we can now assemble a summary of the behavior, as we've come on it here, of the un-needing addict as being:

> meek, "loner," depressed; **not avoiding** falling from high places; or drowning; or ingesting known or suspected poisons; or strangulation; or self injury with tools and weapons; or swallowing in-digestible objects; or harm due to enemies and predators; all this simply continuing to "recur," without "trying," and without "consequence," as the "un-needing" *addict* would have it.

It will therefore happen that, having ceased to avoid those learned behaviors that *advantage* preservation of life, the *un-needing* addict will, not always, but on one occasion or another—by **not avoiding** either falling from high places; or drowning; or ingesting known or suspected poisons; or strangulation; or self injury with tools and weapons; or swallowing in-digestible objects; or harm due to enemies and predators; all this simply continuing to "recur," without "trying," and without "consequence," in a depressed "loner"—enact *learned* behavior that, *not advantaging* life, **dis**advantages it excessively, and, in so doing, become the victim of … suicide. Furthermore, since this is on one occasion or another, even if not always, there will be an identifiable "risk" that the life of the "un-needing" addict will end in suicide.

And there is, indeed, a known (high) risk of suicide occurring among addicts, *opiate addicts*, in particular, who would generally be among the "un-needing"—"loner" and depressed. So such addicts are frequently admitted to hospitals to undergo treatment in the hope of dissuasion. But, in spite of the determined efforts of their caregivers, this is not always successful. Here, briefly, is what has been observed in a group of addicts, in one such hospital:

> Of the thirteen addicts, seven hanged themselves, three cut their throats, two jumped from high places, and one swallowed open safety pins. Five committed suicide in the ward bathrooms, and four did so in their own treatment rooms. One suicide occurred in a seclusion room, and one occurred in the surgical ward of the general hospital section. The remaining two suicides were jumps from high places, one from the roof, and one down a stairwell.

## Serial Killing and Suicide

As we've seen earlier, the bipolar addict can enact behavior which passes between the un-satisfiable (manic) and un-needing (depressive) extremes with an intervening period of normal behavior. However, at the extreme un-satisfiable pole, there can be killing of a human, and this can be repeated each time that the addict returns to the un-satisfiable extreme, leading to "serial killing." But as we have just seen, this same addict could, at the other un-needing extreme, enact learned behavior that, *not advantaging* life, **dis**advantages it, and, in so doing, become the victim of suicide.

It will therefore happen that, not always, but on one occasion or another, as the serial-killer moves through normal behavior to the un-needing extreme, such an addict might enact, from a large repertoire of learned behaviors, behavior that **dis**advantages life, terminating the series of killings in suicide. And since this is on one occasion or another, even if not always, there must be an identifiable "risk" that serial killing will be followed by suicide, or that a suicide will be preceded by serial killing.

However, the general practice, especially in news media, is to report such a series of events as "*a serial killing* ... that happened to have ended with suicide of the killer." Indeed, this is the order in which *I* have just presented it. But the story could equally well, perhaps even better, be made to run the other way, as "*a suicide* ... that happened to have been preceded by a serial killing." Whichever way the story is told, it is a fact that there are, as much recent history reveals, strong links between suicides and serial killings.

## THE EXCEPTIONAL, THE CREATIVE AS SUICIDE-PRONE ADDICT

It is possible to become "exceptional" in the enacting of every kind of learned behavior. So the exceptional can be an exceptional racing-car driver, or violinist, or businessman, or mathematician, or physicist, or writer, or artist, or composer, or preacher, or tennis-player, or financier, and the list goes on. But the exceptional can be even more than this, in *extending* learned behavior itself, in being exceptionally "creative." So the exceptionally creative, or just the "creative," as we say, can enlarge any of those learned behaviors just named. For all this—exceptional, creative, exceptionally creative—I'll be using the one word "exceptional."

How does one become exceptional? Well, to become exceptional, it requires, as the *exceptional* would have it, becoming still better, *even if* the exceptional is already the exceptional best in the world. But, as the foregoing discussion would have it, the exceptional must be *un-satisfiable*—even with being already exceptional—to become *really* exceptional.

Because the behavior of the exceptional is *learned* behavior, becoming exceptional, and, indeed, remaining exceptional entails the pursuit of the particular exceptional learned behavior to the exclusion of at least some other learned behaviors that would, for the "average"— for the non-exceptional, for just the *onlooker*—be "needed." Indeed, as the *onlooker* would have it, the exceptional must "sacrifice," in order to become exceptional; while, for the *exceptional*, all this simply continues to "recur," without "trying," and without "consequence." Thus, at one extreme, the exceptional must be *un-satisfiable*, and, at the other, *un-needing*. What this is saying, then, is that the exceptional must be an addict—a bipolar addict: *un-satisfiable* and *un-needing*; manic-depressive.

But, a little earlier, we have seen the following:

> [T]he *un-needing* addict will, not always, but on one occasion or another—by **not avoiding** either falling from high places; or drowning; or ingesting known or suspected poisons; or strangulation; or self injury with tools and weapons; or swallowing in-digestible objects; or harm due to enemies and predators; all this simply continuing to "recur," without "trying," and without "consequence," in a depressed "loner"—enact *learned* behavior that, *not advantaging* life, **dis**advantages it excessively, and, in so doing, become the victim of … suicide. Furthermore, since this is on one occasion or another, even if not always, there will be an identifiable "risk" that the life of the "un-needing" addict will end in suicide.

There is, indeed, a known risk of suicide occurring among the exceptional, the creative—a higher risk than among the general population. And there is no place, of which I'm aware, in which this risk is made as clear as in the book *Night Falls Fast (understanding suicide)*, written, in 1999, with the recognized authority and engaging grace of Kay Redfield Jamison. Since I refer to it frequently, I've taken the liberty of referring to the book as *NFF*. When a page number is involved, the whole is sometimes written as [*NFF* (number)]. I should add that, although I appeal to the *facts* in *NFF*, it should not be taken that what is being advanced here, as *ideas*, are, or would be shared by Dr. Jamison. Here, then, is some of what *NFF* has to say:

> At least twenty studies have found that highly creative individuals are much more likely than the general population to suffer from depression and manic-depressive illness. Clearly, mood disorders are not required for great accomplishment, and most people

who suffer from mood disorders are not particularly accomplished. But the evidence is compelling that the creative are *disproportionately* affected by these conditions. [*NFF* (180)]

... The toll of suicide on artists, writers, scientists, mathematicians, and others strongly influential on their societies is powerful. The rates of suicide in these groups have been examined in a series of studies conducted by researchers in the United States, Britain, Europe, and Asia. Eminent scientists, composers, and top businessmen were, in these investigations, five times more likely to kill themselves than the general population; writers, especially poets, showed considerably higher rates. [*NFF* (181)]

This is supported by a long list of the exceptional and creative, who have been victims of suicide, including: Ernest Hemingway, Virginia Woolf, William Walton, Ludwig Boltzmann, Alan Turing, Emil Fischer, and many, many others.

The news media are overcrowded with stories of the exceptional lawyer, or golfer, or musician, among other exceptionals, enacting behaviors that involve exceptional indulgence in sexual activity, for instance. And the behaviors often involve what, as the *onlooker* would have it, places the previously "good" reputation of the "exceptional" at risk. So, what does the foregoing have to say about these kinds of behaviors?

The answer can begin with going back a few paragraphs to where it says: "the exceptional must be an addict—a bipolar addict: *unsatisfiable* and *un-needing*; manic-depressive." Then we add to this the following:

> [I]t is "learned" behaviors, in their *unbounded* form, that will become the behaviors of the addict, that, *for the addict*, simply continue to "recur," without "trying," and without "consequence"; that will become the behaviors that appear when, as the neuronal auto-erasure of neurodegenerative disease advances, "satisfaction" and "need" are erased in a brain.

Together, they are saying that the neuronal auto-erasure that, *at the outset*, brought on the particular area of the un-satisfiable which had come to *define* the "exceptional"—in law, or golf, or politics, or music, or finance, or whatever—could, as a *normal* development of the addict's neuro*degenerative* **disease**, have *become* more *widespread* than the auto-erasure that would have defined the particular exceptionality *at the outset*.

This "becoming more widespread" would have given rise to *another* area of learned behaviors in which the addict—the exceptional—had *also* become un-satisfiable. That "the behaviors often involve what, as the *onlooker* would have it, places the previously 'good' reputation of the 'exceptional' at risk," is of no consequence to the *addict*, with behavior, as the *addict* would have it, simply continuing to "recur," without "trying," and without "consequence."

The ramifications of this are wide-ranging, as we can see if we notice that the "exceptional," precisely by virtue of *being* exceptional, will often eventually occupy a position of "power." For if we go again to the observation: "the behaviors often involve what, as the *onlooker* would have it, places the previously 'good' reputation of the 'exceptional' at risk," then this clearly refers to *the-behavior-of-those-exceptionals-that-occupy-positions-of-power* placing their reputations at risk.

And this brings us to the famous opinion of Lord Acton, in his 1887 letter to Bishop Mandel Creighton: "Power tends to corrupt, and absolute power corrupts absolutely. Great men are almost always bad men." (The "bad," with its wagging finger made less generally accusing by the "almost," reminds us that Lord Acton was known to have been a *discerning* moralist.)

As we can now see, this is no coincidence. For, to repeat what the foregoing is saying:

> [T]he neuronal auto-erasure that, at the outset, brought on the area of the un-satisfiable which had come to *define* the 'exceptional'—in law, or golf, or politics, or music, or finance, or whatever—could, as a *normal* development of the addict's neuro*degenerative* **disease**, have *become* more *widespread* than the auto-erasure that would have defined the particular exceptionality *at the outset*. This 'becoming more widespread' would

have given rise to *another* area of learned behaviors in which the addict—the exceptional—had also become un-satisfiable.

Evidently, Lord Acton's years of experience among the "exceptional," wielding "power," had led him to identify what is no less than one of the many manifestations of the neuronal auto-destruction in neuro*degenerative* **disease**, as it advances from a *first* commendable "exceptionality," to some "bad" *next*, tied to "power"—Acton's disease?

## SUICIDE EPIDEMICS

If, as is being advanced in all the foregoing, suicide is learned behavior, it must be that it can be acquired—"caught"—from another human in either or both of the two principal ways that learning can be "caught": listening to someone who had already learned the behavior, and reading about it. What this is saying, then, is that suicide should be able to "spread," as one person "catches" the relevant learned behavior from another. In effect, in some circumstances, suicide should display the features of an "epidemic," as this is known for infectious diseases that one person can "catch" from another.

So, here is some of what *NFF* has to say about this:

> Suicide prevention is not just a clinical problem. Society must deal with the potentially infectious repercussions of suicide, especially among the young, and must somehow try to keep a single tragedy from progressing to deaths of others. The contagious quality of suicide, or the tendency for suicides to occur in clusters, had been observed for centuries and is at least partially responsible for some of the ancient sanctions against the act of suicide. [*NFF* (276)]
>
> ...There has been no shortage of suicide clusters in recent years: they have occurred in psychiatric hospitals and clinics; in suburban America—Plano, Texas; Mankato, Minnesota; Bucks County, Pennsylvania; Fairfax County, Virginia; South Boston; New Jersey; South Dakota—and on college campuses (there were,

for example, six suicides in three months at Michigan State University.) [*NFF* (278)]

## THE SUICIDE BOMBER

"Killing-the-'enemy'" is among our numberless learned behaviors. And even if just *killing* isn't the learned part of that behavior, *what* and *who* is the "enemy" certainly are. What, then, is there about the "enemy" that would have made "killing the 'enemy'" a behavior that became "learned"? The answer is that "killing the 'enemy'" became a learned behavior, in a surviving human species, because the "enemy" is whatever **dis**advantages life, and so "killing the 'enemy'" is one of the learned behaviors that **ad**vantages survival, **ad**vantages life—the killer's life.

So let us see what behavior could become if the neuronal auto-erasure of neurodegenerative disease should lead to an addict that is *un-satisfiable* with respect to the learned behavior: "killing the 'enemy'." This would lead to killing more and more of the enemy. Indeed, no amount of killing the enemy one-by-one would satisfy the un-satisfiable addict. So if the learned behavior included learning how to make and use a mass-killing bomb, then using a bomb, by the un-satisfiable, would become part of behavior. The victim here is the "enemy," killed as a learned act of survival, with the more of the enemy killed, the closer to satisfying the manic un-satisfiable. Interestingly, this is distinctly *unlike* the victim of the serial killer, not an enemy, killed as grandiosity, with *recurring* acts of *single* killing closer to satisfying the manic un-satisfiable.

But if the addict happened to be bipolar, manic-depressive, there would be the other extreme from un-satisfiable, the un-needing extreme, at which, as we have seen above, the addict can become un-needing of even those learned behaviors that *advantage* preservation of life—can become, as the *onlooker* would have it, "sacrificial."

It is not unusual that, given the learned behaviors which would lead to killing the enemy, the killer and the enemy must sometimes be close to each other, and this is still true of even a bomb that an un-satisfiable addict would be able to carry. All this is learned.

It could therefore happen that the un-satisfiable mass-killing bomb-using addict *joins* with the enemy to achieve killing. All this is learned. But, so joined, a bomb that would make mass-killing among the enemy would also kill the addict. All this is learned. And then the two extremes—un-satisfiable and un-needing—of bipolar addiction manifest themselves in an exploded bomb that achieves mass-killing of the enemy, and suicide of the killer. All this had been learned.

And what had been learned could also have been *taught*—by a teacher even *more un-satisfiable* than the now dead addict. Indeed, a teacher so un-satisfiable that the bomb-carrying addict had been merely one more manic-depressive *opportunity* to kill more of the *teacher's* enemy. There are teachers who, more than perceptive, and selective, and wise, are "exceptional."

But if, as *NFF* has just told us, even the old traditional spreaders of learning— listening and reading—can lead to a suicide *epidemic* in the quiet of "suburban America," then what could prevent the *enormous new power* of the world-wide *internet*, as a limitless spreader of learning, from unleashing a suicide-bombing epidemic … of *pandemic* proportions? What a need for caregivers—lots and lots of caregivers!

## THE ERROR IN SEEING "ATTEMPTS" AT SUICIDE

Looking out my kitchen window, I see the neighbor's cat, a usually under-fed tree-climbing beast, chasing a little red squirrel. The squirrel finds a tiny hole in a nearby maple, and dives in. The cat slinks away. That's a failed attempt at catching and eating the little squirrel? Well, I'll never know, since the cat might just have wanted to play with the squirrel. Unlikely? Yes. But I can't be sure, since—in-spite of all the things I'd *so* like to do to the neighbor's cat—getting a confession out of him remains beyond me. (I do like cats, nice cats.)

So, now, I'm in a small store. A man, wearing a face-mask, enters and says to the owner: "Gi'me all the cash you got." The owner reaches down as if to get the cash, comes up with a hand gun, pointed at the man. "Don't shoot", yells the man, who then hurries out the front door. Since I could tell that *this* was a failed attempt, out of sheer curiosity, I followed the man into the street. Seeing that he'd already taken off his mask, I asked him how such a decent-looking fellow, like him, could have been attempting to steal the owner's cash. "Oh," he says, "I wasn't attempting to steal anything at all. I'm actually the owner's cousin, know him well, he knows my voice, and I was just checking to see if he was well prepared to defend himself. *You* were the one I didn't know, that's why I had to make it *look* real." But I wonder if I should have believed him ....

Suppose, now, that we'd seen someone swallowing a handful of pills, someone who months later committed suicide, we might say that *that* had been a "failed attempt" at the future suicide. And if we had seen the same person swallowing *another* handful of pills *a few days later*, we would be *certain* that *that* had been "a failed attempt" at the future suicide.

But we shouldn't be so certain. For, as we've just been seeing:

> **The only way to know what was *very likely* to have been a "failed attempt" is to have *the actor—the very actor* that, according to the *onlooker*, had enacted a "failed attempt"—*tell us, say to us,* if it had *really been* a "failed attempt" ... and, even *then*!**

This is telling us the following:

> **When suicide finally *occurs*, there will *never* be any way of knowing if a preceding act, no matter what it might have been, had *really* been an "attempt at suicide," a "failed attempt"—or an "attempt" at all.**

And we can go further, for this allow us to see why, although "failed attempts" are taken to be among the more "reliable predictors" of suicide, there is a great mass of evidence showing that what the *onlooker* assesses as "a failed attempt at suicide" does not always have suicide as its sequel. Indeed, there is even paradoxical "evidence" that what the *onlooker* assesses as "a failed attempt," although *more frequent* in females than males, has suicide as its sequel far *less frequently* in females than males—except in China! How could a "failed attempt *at suicide*" that occurs *more* frequently in females, have *suicide* as its outcome *less* frequently? What on earth, which could pass for *tenable* explanation, would transform *more* "attempts" into *fewer* "successful" outcomes—except in China?

All of which leads to the following conclusion:

It is merely the **onlooker** who **attributes** to the **victim**—almost always **incorrectly**—"failed attempts" at suicide. And not just with suicide, for there was no "failed attempt" at a "hold-up" in the store. It was I, the *onlooker*, who *attributed*—fully *incorrectly*—to the shop-owner's cousin, a "failed attempt"—signifying no attempt at all! Or, was there? Or did I?

The path to suicide doesn't lead through a forest treed with "attempts" at suicide. Rather, it leads through a dust-stormy desert touched by neither "attempts" nor "trying"—just "happenings." It would be useful to repeat, right here, an earlier paragraph that speaks to this:

It will therefore happen that, having ceased to avoid those learned behaviors that *advantage* preservation of life, the *un-needing* addict will, not always, but on one occasion or another—by **not avoiding** either falling from high places; or drowning; or ingesting known or suspected poisons; or strangulation; or self injury with tools and weapons; or swallowing in-digestible objects; or harm due to enemies and predators; all this simply continuing to "recur," without "trying," and without "consequence," in a depressed "loner"—enact *learned* behavior that, *not advantaging* life, **dis**advantages it excessively, and, in so doing, become the victim of … suicide. Furthermore, since this is on one occasion or another, even if not always, there will be an identifiable "risk" that the life of the "un-needing" addict will end in suicide.

And there really *is* a "risk," which is to say no more than that "the *un-needing* addict will, not always, but on one occasion or another … enact *learned* behavior that, *not advantaging* life, **dis**advantages it excessively, and, in so doing, become the victim of … suicide." But that's *all* we can ever know from "attempts"—"a risk." The question of *which act* was an "attempt at suicide," and which was not, is destined to be awaiting an answer, up to the very end—an answer that can never come.

But might there be a predictor of ultimate suicide less capricious than "failed attempts"?

## "Hopelessness" the Predictor of Suicide

What, then, is this "hope" that must depart, so "hopelessness" can take its place? "Hope" is aimed at "the future," for it is "the future" alone, not the present, that can carry what "hope" would bring. The present is already too far gone for "hope" to leave its mark. So, for "hope" to be more than merely a sound signifying nothing, there must be, flowing among the quiet "as ifs" of a brain's neurons, nowhere else, that persistent ghost of a "yet-to-come"—"the future."

But this is "the future" about which, as shown earlier: "we can see why neurodegenerative disease will, as one of its manifestations, erase 'the future'," summarized in the sequence:

> neurodegenerative disease → neuronal auto-erasure → erasure of "the future."

Which now allows us to see that neurodegenerative disease will, in erasing "the future" as one of its manifestations, erase "hope" as well—allowing "hopelessness" to take its then empty, futureless place, as summarized in the following sequence:

> neurodegenerative disease → neuronal auto-erasure → erasure of "the future" → erasure of "hope" → "hopelessness."

This way we can see why, when "hopelessness" sets in, its host arrives at a "place"—a now degeneratively *diseased,* "futureless" place—from which there is no further "place" to go; no further "place" to "hope" to go. Indeed, no further place even to "need" to go. So "hopelessness" must mark extreme "un-needing"—the extreme condition of the un-needing, loner, addict.

To see what this implies, let us return to the following earlier paragraph:

It will therefore happen that, having ceased to avoid those learned behaviors that *advantage* preservation of life, the *un-needing* addict will, not always, but on one occasion or another ... enact *learned* behavior that, *not advantaging* life, **dis**advantages it excessively, and, in so doing, become the victim of ... suicide. Furthermore, since this is "on one occasion or another," even if not always, there will be an identifiable "risk" that the life of the "un-needing" addict will end in suicide.

So, with this, we can see why "hopelessness" is an *almost unfailing predictor* of *ultimate* suicide—far beyond just "risk," now. Because, for the ultimate outcome to be any different from suicide, the depressed addict would have to be "rescued" from the *un-needing* state of "hopelessness"—would require that the already *existing* future-erasing, by neuronal auto-destruction, be *reversed*. But this is the *very opposite* of what will tend to happen—even with the devotion of expert, determined caregivers—since the *neuronal auto-destruction* that *entailed* "hopelessness" will, *degeneratively*, worsen as it worsens, rendering ever deeper the then inescapable *futureless* state of "hopelessness," with its relentless drift, one way or another, far beyond just "risk," now ... into the learned abyss of suicide—as certain as "certain" can be.

Here is how all this looks to the clinician:

> In short, when people are suicidal, their thinking is paralyzed, their options appear spare or nonexistent, their mood is despairing, and hopelessness permeates their entire mental domain. The future cannot be separated from the present, and the present is painful beyond solace. [*NFF* (93)]

"Hopelessness" is the expression of an *existing* neurodegenerative *disease*, incurable as is all such disease, becoming *fatal* in the particularly sudden, learned way we know as "suicide."

Is a person's degree of "hopelessness" measurable—diagnosable? A substantial amount of research has been done on that question, with a strongly affirmative answer. This has formed a crucial result of the research on the relationship between hopelessness and ultimate suicide. Notable among the researchers has been Dr. Aaron T. Beck, leading to the "Beck Hopelessness Scale," as a sensitive indicator of suicide potential.

What, in evolutionary development, could make suicide—so evidently opposed to survival—an inseparable part of the behavior of members of a surviving human species? The answer is simply that suicide, *in itself,* has *no* survival value. However, as just another expression of neurodegenerative disease, suicide derives *its* "inexplicable" presence from the presence of neurodegenerative disease *itself,* which derives its *own* "inexplicable" presence from the evolutionary departure that bestowed, on the earliest brain, neuronal auto-destruction as the path to "survival," eons upon eons ago. This would have made of neurodegenerative disease, and suicide within it, an eternal evolutionary paradox—survivalist auto-destruction!—but for there being no place for paradox in evolution's unforgiving space, only resolution coming in the un-ending challenge of explanation.

## THE FATTEST, THE THINNEST, AND THE SUICIDAL

The Swedes, record-keepers unequalled of population statistics, are exceptionally good at diligent epidemiology that confounds "common sense." For 31 years, Swedish researchers followed up 1.3 million male military recruits only to find that the rate of suicide was *highest* among the *thinnest*—not the *fattest,* as "common sense" would cast belief.

And "common sense" would so cast belief because, still according to "common sense," it is the *fattest* who will, because of their evident "condition," be the more likely to be, somehow, "sick of life"; depressed; embarrassed by what others (the thinnest, presumably) will "think" of them—making them the easier targets of suicide. But the foregoing explanation of suicide tells us why what the Swedish researchers are saying trumps "common sense."

We can begin to see why by noticing that, as shown earlier, it is the **un-satisfiable** who will become the **fattest**. It is the inability of any amount of food, short of (and maybe even in spite of) vomiting, to *satisfy,* which sustains the eating that leads to the fattest. However, as also shown earlier, it is **not** at the extreme of the *un-satisfiable,* except for the ultimate extreme of the "exceptional," that *suicide* is at all frequent—serial killing, perhaps, but not suicide.

What ushers in suicide is the version of neurodegenerative disease, with its neuronal auto-destruction, that brings on the *other extreme* from the *un-satisfiable* eater—the version expressed in the **un-needing**, giving rise to *anorexic* **thinnest**—who, at the *un-needing* extreme, slides, as just shown, into "hopelessness," with almost un-failing suicide.

Here, then, is how the connection between the *thinnest* and their exceptionally high rate of suicide comes about:

> It is the *same* extreme *un-needing* that leads to "hopelessness" and *suicide*—arising in the neural auto-destruction of neurodegenerative disease—that leads *also* to the *anorexic* **thinnest,** which *connects* the **thinnest** to a *much higher rate of suicide* than that of the *un-satisfiable* **fattest**.

The Swedes, record-keepers unequalled of population statistics, are exceptionally good at diligent epidemiology that confounds "common sense."

## SUICIDE AND THE SEASONS

There is conclusive evidence that the *rate of suicide* varies with the *seasons*. The highest rate occurs in the *summer*, in both the northern and southern hemispheres—in the *local* summer. A particularly full reference to this begins at [*NFF* (206)]:

> The seasonal variation in suicide is one of the most robust and consistent findings in the research literature. In the late 1800s, Enrico Morselli studied suicide in eighteen European countries and showed that in seventeen of them the maximum suicide rate occurred in the spring or summer months. (Conversely, in virtually every country, the minimum occurred in the winter.) Several years ago, I [Jamison] reviewed more than sixty studies of seasonal patterns in suicide and found a similar pattern. ... Consistent with this, studies carried out in the Southern Hemisphere—in Australia, Chile, Uruguay, and South Africa—show that suicide peaks in the months making up their springs and summers.

Why, then, should suicide attain it maximum rate in the summer? The question is particularly interesting because "common sense" would suggest that, if suicide is to be seasonal at all, it would have its *highest* rate in *winter*, when human "mood" might match that season's darkness. But what we will see is that this seasonality is almost certainly connected to the seasonality of *sleep*, more particularly, the seasonality of *disorders in sleep*. Why would that be?

To answer the question, it will be convenient to begin by recalling the relation between *disorders in sleep* and *neurodegenerative disease*. So let us return to Chapter V for the following:

> It becomes evident, from all this [preceding discussion of disorders in sleep], that there is, and always has been, *an evolutionary connection* between **disorders in sleep**—amounting to its quasi-absence—*and the development of a destructive **neurodegenerative disease***.

What this is telling us is that, if there are seasonal variations in "disorders in sleep," there will be seasonal variations in neurodegenerative disease. And we have seen, in an earlier section (on "hopelessness") that neurodegenerative disease will *entail* "hopelessness." But the best available clinical observation confirms that "hopelessness" is an almost *invariable prelude to suicide*.

So, in summary, we have this sequence:

> disorders in sleep → neurodegenerative disease → hopelessness → suicide.

From which we can have as conclusion: If there are seasonal variations in the rate of disorders in sleep, there will be the *same* seasonal variations in the rate of suicide. So, next question: Are there seasonal variations in the rate of disorders in sleep?

We can begin the answer by noticing that if there are to be seasonal variations in the rate of disorders in sleep, the *highest* rate will tend to occur in *summer*, when the *strong* sunlight of *longer days* would interfere with sleep—would *favor* disorders in sleep. Putting this the other way round: If there are to be seasonal variations in the rate of disorders in sleep, the *lowest* rate will tend to occur in *winter*, when the *deep* darkness of *longer nights* would encourage *un*-disturbed sleep—would *inhibit* disorders in sleep.

Returning to the sequence above, we would then have that the tendency, just suggested—for the *highest rates of disorders in sleep* to occur in *summer*—would lead to a tendency for the *highest rates of suicide* to occur in *summer*, which is in agreement with what is given in *NFF*, as the summer-peaked seasonality of suicide.

Evidently, what would help to confirm the present argument is *research-evidence* showing that there is, indeed, a higher rate of *sleep disorders* in summer than winter. The most relevant evidence, that I've been able to find, comes from research done in Finland, on the seasonal variation of the rate of sleep-related motor accidents. Part of the conclusion is as follows: "The results showed a marked seasonal variation such that the share of sleep-related cases [of accidents] was much higher in summertime." This is the kind of evidence that's needed. And it is particularly valuable coming from a high-latitude country like Finland, where there are marked changes in daylight between summer and winter.

## BACK TO "THE GENERAL ANSWER" AND "MENTAL ILLNESS"

All the behaviors discussed in this chapter have been explained as being, one way or another, manifestations of neurodegenerative disease—the same neurodegenerative disease generally known, and shown in earlier chapters, to underlie Alzheimer's and Parkinson's diseases. Is there, then, some *single* conclusion, arrived at somewhere in an earlier chapter, that would account for this striking generality? That conclusion, reached in Chapter VI, and repeated now, is what was named "the General Answer":

> Neurodegenerative disease involves the degenerative auto-destruction of the neuronal components that make up a brain. Therefore, as the degeneration progresses, **all the functions of a brain, whether *"sensorial,"* or *"mental,"* or *"directorial,"* could eventually fail.** "Sensorial" functions are those associated with the brain's involvement in sensation, as in hearing, taste, touch, smell, and sight. "Mental"

functions are those associated with brain-originated manifestations in the brain's host such as memory, imagination, "standing for," "as if," and so on. "Directorial" functions are those associated with the brain's direction of parts of its host: legs, hands, feet, tongue, for example.

So the present chapter started becoming what it is when, having written "the General Answer," I stopped to notice that "the future" is "mental"; "consequence" is "mental"; "satisfaction" is "mental"; "need" is "mental"; "attempting" and "trying" are "mental"; "hope" is "mental"—all "mental," like the "mental" of "the General Answer," in that they must reside nowhere but in the "as ifs" of a brain's neurons. Then, given what "the General Answer" was saying, it became increasingly clear that neurodegenerative disease's auto-destruction of the neuronal components which make up the "mental" functions of a brain, must manifest itself in ways, "diseased" ways, not *yet* regarded as the expressions of such auto-destruction at work.

It was in this gradual, sometimes halting way that what are generally taken to be evidence of "mental illness"—addiction, homelessness, serial killing, depression, hopelessness, suicide—have here been brought within the *single* framework of neurodegenerative disease, beside Alzheimer's, and Parkinson's and all the other well-known expressions of that disease. And, in so doing, they have been brought within the general *necessity* that, for a brain to be free of "mental illness," as for a brain to escape neurodegenerative disease, a brain *must sleep*.

The message from "the General Answer" is therefore that "mental illness" is the expression of the neurons of a brain engaged in their own degenerative destruction, with the consequent destruction of the "as ifs" that only neurons can host. And it is the *nature* of the "as ifs" themselves—"future," "consequence," "satisfaction," "need," "attempting," "trying," "hope," all *phantoms* in their way, all *make-believe*, that support "normalcy" and "mental health"!—as they are increasingly disrupted by neuronal auto-destruction, that gives "mental illness" *its nature*, as somehow not quite based in the cellular firmness that alone evolution must work with; indeed, as somehow meta-bodily "mental."

But then we are reminded, as evolutionary explanation always brings us to be reminded, of the *hereditary* character of "mental illness," now, as this chapter is saying, tied to the *hereditary* nature of neurodegenerative disease, as it is *itself* tied to the inescapable *hereditary* nature of at least some *disorders in sleep.*

However, there is still one "as if" that is grander and more peculiarly human than any other. Let us turn to that now.

# CHAPTER VIII
# NEURODEGENERATIVE
# DISEASE AND
# SCHIZOPHRENIA

## ABOUT THIS CHAPTER

It was a really cold Saturday morning—minus twenty Celsius. So we were glad to turn into the little local library, nice and warm. There, in what is usually the reading room, was a corner filled with children, between about five and seven years old, enthralled into silence by the lady telling a story. We waited quietly to the end of the story to hear, amid the clapping, high-pitched cries for more!

Well, not a word of the lady's story was true—all imaginary modern stuff, not even as good as an old-fashioned fairly tale—must have come straight out of her brain ... like this little story that *I've* been telling: not true, just came out of my brain.

Look's like he's already forgotten the title he gave the chapter—something about schizophrenia. What does all this about "story telling" have to do with schizophrenia? And how come he's now talking about *himself* as "he," rather than "I"? All this looks a bit as if ....

Forgotten? No, he hasn't forgotten, because, as he's going to show in this chapter, schizophrenia has *almost everything* to do with story telling. And it's because he's committed to *evolutionary* explanation that *how* we could ever have *become* story-tellers, in the first place—how humans could ever have *become* the unique vehicles for schizophrenia—looms so large.

Just think of it. Stories are full of characters that come straight out of the brain of the story teller—nothing real about them at all, just *words*. To save them from being outright fibs they need an introductory disclaimer, like "Once upon a time." Compare this with the situation in which you and I are face to face, in the **presence** of each other. For me, you are *not then* just "words," you are what we call "real," in that I can see, or hear, or touch you. Then I can learn behavior *directly* from you, behavior that I can use to increase my chances of survival—how to hunt, or dance.

So let us go back now to the characters in *my* little story, about the lady telling *another* story. *You* can never be face to face with them; you can never learn behavior *directly* from them; for you, they will never have the **presence** that can exist between us, they will always be just *words* that fade away; they will always be "not there with you," they will always be **absent**.

But, even if it's all just words, story-telling is an integral part of the behavior of all members of the surviving human species. There are whole piles of them held in brains, and even more of them in books. So, at some point, evolution must have added to the behavior of emerging humans the capacity to tell stories that had no *real*, *live* characters in them at all, just *words*. Furthermore, that *natural selection* didn't send evolution's new story-telling addition, to emerging human behavior, into the evolutionary trash heap, tells us that there must be *something special* about story telling that would let it get by such an unforgiving natural strainer.

But even supposing that there *is* such a "something special" (and as we'll see, later, there certainly is), there is still a big question that an evolutionary explanation must answer. The question goes like this: What good would it be, for *your* survival, to clutter up *your* brain with *my* stories, if *you* couldn't turn the *word-characters* in *my* stories—good as *dead*, as they happen to be—into some way of affecting *your* behavior, and *your* survival? The answer is that, if evolution hadn't endowed humans with *some* way of causing the actors in stories—the lifeless phantoms housed in words—to enact **behavior** that would affect their **survival**, then *natural selection* would have packed off the carriers of a brain, with such useless stuff burdening it, to the evolutionary trash heap.

So the question isn't *whether* evolution endowed humans with a way of deriving behavior from the characters that came and went in just words. Instead the question is: With *what* did evolution endow emerging humans that would have allowed them to derive behavior from the characters that came and went in the words of stories? What do we do, as the continuing favor of natural selection would have required that we do, to *transfigure* the evanescent verbal phantom in a story into a source of behavior that can affect survival?

The answer, as we'll see, resides in the relation between the "meaning" of a "word" and the **behavior** that the "meaning" must bring on. And, as we'll also see, it is that relation which makes of humans the unique vehicles for schizophrenia. Let's see now how that comes to be.

## THE MEANING OF "MEANING"

Suppose we are standing together and I say: "run." Then there would be some particular behavior that I am expecting you to perform: the behavior which, according to *me*, "run" is supposed to bring on, in *you*. If you then perform the behavior which, according to me, "run" is supposed to bring on in you, I would say "run" has a "meaning" for you—a "meaning" that, as your **behavior** implies, is the *same* for you as it is for me. This is telling us the following: "Run" has a "meaning" which is the *particular* **behavior** that "run," when uttered, brings on in you and me.

There are thousands of sounds like "run"—"words," as we call them—that are an integral part of the behavior of members of the surviving human species (except for that minority among us who, for one reason or another, lack the ability to make use of them). "Words" only need to be uttered, and so are integral parts of the behavior of those members of our species who, even now, neither write nor read. Thus, as all the available evidence suggests, "words" must have been part of what natural selection let ride, as evolution added them to the changing behavior of members of an emerging human species.

Which raises the following question: What could it be, about the evolutionary emergence of "words," that *natural selection* would have *favored* in the ongoing survival of the emerging human species? The answer must reside in some *survival advantage* that uttered-"words" provided for emerging humans. And we can readily see what that survival advantage must have been, with uttered "words" having their "meanings" in **behavior**. Because darkness, dense bush, corners around rocks, for instance, which would **block** *visual* contact, would **let pass** the *uttered* word—an uttered "quiet," with its meaning in **behavior**, making the difference between survival and death for an audibly fretful child, then beyond being *seen*.

Without a "word" having its "meaning" in the **behavior** of emerging humans, there could have been *no evolutionary outcome*, related to "words," on which *natural selection* could have passed its favorable judgment—still so clearly evident, as right here, in today's wordy members of the human species.

## Grammar's "Persons": First, Second, and Third

Now, there is something I need to call to your attention: everything in the previous paragraphs, about "words," is going on between *you* and *me*, even as it continues here. I have been assuming that we are "together," so that the "I" is the "me" that *you* could *see*, or *touch*, or *hear*; and the "You" is the "you" that *I* could *see*, or *touch*, or *hear*. And this has something to do with "grammar"—yes, grammar, that seemingly useless torture in which early teaching delights. But grammar isn't quite as useless as it might seem, because it now draws our attention to what is referred to as the *"first* person," the "I"; and the *"second* person," the "You." So, grammar has things organized such that, if we are together, what goes on between us runs in the first and second persons—"I," "me," "we"; and "You." (There used to be a "thou" and a "thee" in English, now happily banished.)

But let us now suppose that, although we are together, we want to refer to something that is neither You nor I; and let it be something that neither You nor I can *see*, or *touch*, or *hear*. Well, grammar is all prepared for that too, with what is known, appropriately enough, as the *"third* person"—"he," "she," "it"; "they," "them." Then grammar goes even further with so-called "nouns," "names"—cat, elephant, John, Napoleon—not just pro-nouns, or kind-of nouns, like "I" and "You."

However, there's a big problem with this *third* person in grammar—a big problem for evolutionary explanation, especially. And we tend *not* to notice it, because we've become so accustomed to, and expert in using grammar's *third* person.

To see what the problem is, we begin by recalling that, with the *first* and *second* persons, natural selection would have allowed uttered "words" to become an ongoing feature of members of a surviving human species because "words" have "meaning," and "meaning" is expressed in **behavior**. So, even for the child who can't be *seen*, but *known* to be *really* there, the "word" "quiet" implies the **behavior** of a "you," which the "meaning" of "quiet" should engender.

And it is this something that is *really* there—"I," and "You," for instance—that so characterizes the *first* and *second* persons, which highlights the problem with grammar's *third* person. To begin seeing why, look at the following use of the *third* person in this line:

*She lives behind the rock.*

Then compare that with the following use of the *first* and *second* persons, in this line:

*You and I run together.*

What's the difference between them that I now want to stress? Well, beginning with the *second line*, the "I" who does the uttering is in the **presence** of the "You." So there is no problem in how what is uttered could translate into **behavior**. For the "I" that does the uttering, the "You" is *real*—can be *seen, touched, heard*—and "run" has a "meaning": in the kind of **behavior** it requires, which the *utterer* and *hearer* share. There is a **presence** existing between the "I" and the "You" which makes "I," the utterer, and "You" the *hearer* "real" for each other.

But this is far from the case with the *first line*. For instance, the *third*-person "*She*" of the *first* line has **no necessary presence,** so far as the **hearer** is concerned. Indeed, there might not have ever **been** a "*She*," even for the *utterer* of the *first line*. And it is this possible **complete non-presence**—possible **complete absence**—of the *third* person which poses a challenge for evolutionary explanation. Because, since, as evolutionary explanation would have it, it is the "meaning" of "words" as **behavior** which gives human survival value to the uttering of "words," there is the question of how there could be **behavior** that could attach *any* survival value to the uttering of *third*-person "words," which could be just noises, without any necessary **behavior and meaning.** They could be just noises signifying no more than what *has never even existed* except as *fictions* in the "as ifs" of the brain of the utterer—just ghosts, lifeless phantoms that lack essential behavior, and reason for survival.

This implies the following:

> For the *third person* in grammar to have become, as it has become, an *integral* part of the behavior of members of a surviving human species, the way in which a brain makes use of grammar's *third* person, which can be **completely absent**—have never existed; good as dead—**must** be *fundamentally different* from the way it makes use of the *first* and *second* persons, which are **always present**, actually **existing**.

And this brings us to a truly challenging question: What must have happened in the evolution of how a brain makes use of grammar's uttered *third* person—ghost, lifeless phantom that lacks essential behavior, and reason for survival—so that it could have become, as it has, an integral part of the behavior of members of the human species?

We can move toward the answer by noticing that, for grammar's uttered *third* person to have become, as it has, an *integral* part of the behavior of members of the human species, the *hearer* of the utterance must have become able to *transfigure* the **absent** *third* person, in the utterance, into a source of **behavior** that the *hearer* could **enact,** and secure **survival benefit** from—the *hearer* must be able to so *transfigure* the uttered third person as to make it **behave**. Without the **absent** *third* person being *transfigured* into such a source of **behavior**, there could have been no **behavior**-related *survival value,* for the *hearer* (or the *utterer*), on which *natural selection* could have arrived at its still favorable verdict.

This brings us, now, to the answer, which is as follows:

> In the evolution of how a brain makes use of grammar's uttered *third* person, so that it could have become, as it has, an integral part of the behavior of members of the human species, the *hearer* of grammar's uttered *third* person must have become the **hereditary** carrier of the ability to perform the following:

(a) Admit the *third* person into its brain "as if" the *third* person had been a *second* person—a "You"— and **present**;

(b) **Endow** the *third* person with such "meaning" and the meaning's **behavior** as the **hearer** can endow it with—make the *third* person, **behave;**

(c) Enact, as the situation warrants, the **endowed** behavior of the *third* person, thereby directing the *hearer's own behavior*, in a way that can affect the *hearer's* survival.

In what follows, I'll often use the word "narration" for what I've been calling "storytelling."

## THE BENEFITS OF STORYTELLING

Fortunately, the benefits of storytelling are far easier to explain than the processes that led to and support it. For what it confers on the emerging human is the capacity to pour into the brain of each member of its species, by word of mouth, significant portions of the experience of all the other members with whom it comes in contact, and even other members *beyond* that. Consequently, not only does the behavioral repertoire of each pre-human show its release from instinct as it was enriched by its own increasingly complex "I-You" experience, but now, by means of narration, this repertoire is enriched even further by the shared experience of these new storytellers that begin to appear in increasing numbers.

Prior to the emergence of narration, the behavior of each pre-human was limited to the exploitation of second-hand experience, of mixed vintage, stored in its genes, augmented by its *own* first-hand experience gathered in direct experience. However, narration provided *new* sources of second-hand experience, all now of *recent* vintage, on which to draw, for tuning behavior to its *current* environment. Also, except for the mixing that occurs in mating, the experience stored in genes is loaded into the host in *series*, generation after generation. But, beginning with learning from direct experience, and increasing vastly with narration, experience and related behavior are loaded into the hearer in *parallel*, from like creatures around it, and even from those *beyond*.

Narration must, therefore, have given rise to explosive benefits for our ancestors, so much so that one wonders whether the rate at which the rest of our biology could evolve has ever succeeded in coping with the enormously increased rate of intake of experience that narration made possible. This becomes an even more pressing wonder when we realize that *third*-person animation would not have been limited to the *humanoid* actors alone, in narration, but extended to "animals" and "things" as well, as I shall explain a little later on.

## NARRATION, HUMANNESS AND SCHIZOPHRENIA

As we can see, the emergence of narration must have been accompanied by a seminal transformation in the relation of "humans" to their environment. I believe that this was the final transformation during which the human species emerged. In order to lend more credence to this position, I shall discuss a number of concomitants of narration that have clear expression in humans, and which could not have existed, particularly *together*, prior to the emergence of narration. The concomitants of narration to be discussed are the following:

- Grammar;

- Selfness;

- Being Led;

- Narration and Myths;

- Empathy;

- Untruth;

- Unreality;

- Schizophrenia.

Schizophrenia is included in the list because, as I shall show, it is almost certainly a "mental illness" associated with failures, brought on by the neuronal auto-destruction of neurodegenerative disease, in the area of the brain's neural system that supports narration—indeed, that supports our final humanness. In order to facilitate the demonstration of this, the discussion of each concomitant listed will include a sketch of what would be expected to be the effect on behavior of a *failure* in the neuronal system supporting narration, as would be brought on by the neuronal auto-destruction of neurodegenerative disease. These sketches of the effect on behavior of a *failure* in the neuronal system supporting narration will then be compared with the known symptoms of schizophrenia.

## NARRATION AND GRAMMAR

There has already been considerable discussion of the relation of grammar to narration. Given that, all I'll be doing here is look briefly at how grammar's *third* person might have emerged in an evolutionary sequence, as well as a possible consequence of a failure in the brain's neuronal system that supports narration.

So, imagine an "I-You" encounter between two pre-human siblings, between whom the *first* and *second* person sufficed for the necessary communication. It would have been easy for expansion of the encounter to begin with nothing more complicated than an encounter between *three* siblings. Then, as soon as the "other" person involved in the encounter moved *beyond* the *immediate* I-YOU context of an encounter between just *two*, further engagement with the "other" person could continue with *gestures*, of which *pointing* would almost certainly be the most useful and available. A **third** "person" would have found its way into what is being communicated, albeit with a *gestural* basis for its inclusion.

But though such a *third* "person," increasingly distant, could still be experienced directly, this would no longer be in the immediate I-YOU context that is characteristic of an encounter between just *two* pre-humans. Thus, there begins a slow drifting of the subjects of an I-You encounter from immediate actors, to those more and more distant, until *pointing itself* fails. In the final stage of this slow evolutionary process, the *third* "person" becomes an entirely absent phantom, no longer addressable even by pointing, indeed, a phantom inaccessible to any direct sensing, and which can be conveyed only in the full indirectness of surrogates for senses assembled, in a gradual evolutionary process, from uttered *words*. The ultimate form of the actor in narration then becomes that of a *word-borne* phantom, totally without any *direct* existence, so far as hearers are concerned, and able to affect their survival only to the extent that each of them can endow such a word-borne phantom with some sort of "learnable behavior," by *animation*.

There is no place, other than the "as ifs" of *neurons* in the brain of a *human* hearer and speaker of grammar's *third* person, in which such a word-borne phantom could have been born, bred, and kept "alive."

Suppose, therefore, that the neuronal auto-destruction of neurodegenerative disease should now find its way to the *neuronal* "as ifs" in the brain of a *human*—for whom the use of grammar's *third* person had heretofore been normal—in which the word-borne phantoms of narration must reside. This would limit the victim to the use of the *first* and *second* person only, thus producing the effect of an illness distinguished by a *major loss of normal human speech*.

## NARRATION AND SELFNESS

The uttering of "I" conveys to a "You" a certain "I-ness" appropriated by the utterer—an I-ness that we usually refer to as a "sense of self," or just "selfness." A concomitant of this I-ness of the utterer is the "feeling" of being a "separate" "I," with "separateness" sufficient to define an "inside" and an "outside" of the "I"—the "outside" being "all the rest," including any "You," from which the utterer, the "I," "feels" *separate*.

But there is more to selfness than just "feeling separate," with an *inside* and an *outside* that goes with it. For it is from this "inside" that the "I" "feels" that its **behavior** is "directed." And, indeed, it would seem "normal"—reasonable, explainable, "sane"—for the "I" to feel that *its* behavior is "directed" from *its* "inside," especially when, as with much of the behavior of the "I," it actually does originate in *its* brain, which would clearly be a part of "inside." All this is the same for a "You" when, as will happen, it occupies the reciprocal role of "I."

However, with narration, the process leading to **behavior** in the *hearer*—compared to behavior originating and running between "I" and "You"—is more complex, with more stages, involving as it does, the *hearer's* necessary *transfiguring* of the *third*-person, *word*-borne phantoms that appear in narration. And although this must also be accompanied *normally* by the retention of the feeling of *selfness*—with behavior, originating even in narration, still "coming from *inside*"—unavoidably, there will be ways in which the process associated with narration might fail, ways associated with the complexity, with the stages, of the process itself.

There is no place, other than the "as ifs" of *neurons* in the brains of *human* hearers and speakers of grammar's *third* person, in which the complex transfiguration of a *third*-person phantom into **inside-originating** behavior of an "I"—a *self*—could take place.

Suppose, therefore, that the neuronal auto-destruction of neurodegenerative disease should attack those "as ifs" of *neurons* in the brain of a *hearer* in which the complex transfiguration of a *third*-person phantom, into **inside**-**originating** behavior of a *self*, must take place—a hearer for whom the use of grammar's *third* person had heretofore been normal.

Then we can expect cases of "illness" arising in which victim-hearers will report a **mix** in their **directors** of behavior: some, simply as normal, directing behavior from "inside"; but also some directing behavior from "**outside**," with "actors," which originate in *narration*, doing the directing from "outside."

In the extreme case of this, directing of behavior can move **entirely** to the "**outside**," then victim-hearers present themselves *as if* they *were* the actors. Such behavior would then be said to show signs of "delusion."

Thus, the neuronal auto-destruction of neurodegenerative disease can lead to an "illness" in which behavior is characterized by delusion, involving actors derived from *narration*. But since "normal" humans display a singular, persistent "*inside-originating* self," delusion involving the presentation of a new and different "*outside-originating* self" is very striking, and tends to dominate what are taken to be the indicators of such illness. As we will see, because of what narration can involve, these "outside actors" can take a wide variety of forms.

## NARRATION, BEING LED, AND "DO AS I SAY"

There can be a case of narration in which the *narrator* recites a story in which the *narrator* is an *actor*, giving *instructions* on how to actually *derive* behavior from the *actor-narrator*. For instance, the narrator might say: "When we go to see the dead King, you salute him, as I will."

This is a very interesting case of narration because, in it, the *narrator*, by placing the story in the *never-is-here* "future," is, in effect, making of the *narrator*, in the story, a "word-borne phantom"—just the way that grammar's *third* person is a word-borne phantom that "never-is-here"—with, as we'll now see, a very significant outcome.

And the outcome is very significant because, when the *hearer*—the "You"—treats this narration in *the only way it can*, which is as evolution and natural selection have written the hereditary prescription—transfiguring the quasi-*third*-person, the "in-the-future" "I" into a source of behavior—the *hearer* will have *adopted* the *narrator* as an *originator* of at least some of the *hearer's* behavior, as directed, in the normal way, from the hearer's "inside."

In such a case, the *narrator* could be said to be able to "lead" the *hearer*, that is, to function as a "leader," and exercise at least some degree of "control" over the "future" behavior of the *hearer*. The *narrator* would have come on a way of "talking behavior into the hearer." The narrator would have come on the imperative, albeit initially weak, of "do as I say."

Because of the importance of this development, it will be useful to look again at how it comes about. To achieve "talking behavior into the hearer," the *narrator* needs to place the story, involving the *narrator*, in "the future," because of the inevitably *absent, intangible, phantom-nature* of "the future." For it is by the *narrator* placing the story in this way that the *hearer* is *constrained*—by the hereditary process arrived at by evolution and natural selection, for deriving behavior from an absent, *third*-person, word-borne phantom—to treat the *narrator* in the story as it would *normally* treat an actor in a narration, from which it *derives* behavior, by transfiguration. By situating **him-** or **herself** in "**the future**," the **narrator** becomes the *quasi phantom* which the **hearer** has, normally, *only **one way*** to treat the narrator: **adopt the behavioral direction of the narrator**.

The limiting version of this kind of story is that in which the *narrator* is *also* the *hearer*, that is, recites to him- or herself about him- or herself *alone*, in "the future," as in: "When I sit down to dinner, I will eat only what is put first on my plate—no more." In such a situation, the narrator-listener would be able to achieve a degree of regenerative auto-control of behavior by narration; which could be emphasized by further narration, and so on.

Evidently, we have here come on the fact that the emergence of narration is inseparable from the emergence of the ability of a *narrator* to control the behavior of a *hearer* by means of storytelling, and "suggestion." Thus, the narrating human is, of necessity, the "suggestible" human.

There is no place, other than the "as ifs" of *neurons* in the brains of *human* hearers and speakers of grammar's *third* person, in which the complex transfiguration of a quasi *third*-person phantom, set in "the future," could take place.

Suppose, therefore, that the neuronal auto-destruction of neurodegenerative disease should attack those "as ifs" of *neurons* in the brain of a *hearer* in which the complex transfiguration of a *narrator*—presented as a quasi *third*-person phantom, set in "the future," thereby *controlling* the *hearer's* behavior—would normally take place.

Then we can expect a case of "illness" arising in which the victim-hearer will be unable to follow *suggestions* as these would arise in *narration*, and the person would appear to be *stubborn* and *incorrigible*. It would not be possible to "talk behavior into the victim-hearer." In addition, it would be difficult or impossible to induce hypnosis in such a person.

## NARRATION, MYTHS, UNDERSTANDING AND MYSTERY

It will be convenient to distinguish here between a "phenomenon" and an "entity." So, by "entity" I'll mean something like "sun," and, by "phenomenon," something like "sunrise"; that is, "entity" refers to a more or less separable item in the environment, and "phenomenon" to an episodic sequence in it. Clearly, this distinction is one of convenience in explanation only, since we can find a phenomenon within a phenomenon, and then the "inner" phenomenon will behave like what I have called an "entity."

Now, as the benefits of story-telling are expressed through the expanding behavior of narrating humans—tellers and hearers alike—tuned by the filtering of natural selection, the possibility arises of narrating humans acquiring the behavior of a *phenomenon*, as distinct from that of its simpler entities—like "You," in the earlier cases. A way of accomplishing this, which eventually falls within the behavioral range of the humans that have begun the process of narration with its necessary animation, is for the narration *itself* to include the animation of the phenomenon—the *narrator* does the animating—and so reduces the process that would be needed in the *hearer*, to do the animating.

Thus, in the case of *sunrise*, the narration itself can contain the story of an "actor" that rises early and drives a flaming chariot across the sky. The *hearer* treats this narration in *the only way he/she can*, which is as evolution and natural selection have written the hereditary prescription—transfiguring the quasi-*third*-person "charioteer" into a source of acquired behavior. The acquired behavior then consists of the *hearer* simply *repeating* the narrator's story of the "charioteer"—saying it over and over again to other hearers. When this happens, we say the story, charioteer and all, has become the "understanding" of the phenomenon of sunrise—has become the way the hearers have come to "understand" the phenomenon.

Evidently, the version of the phenomenon provided in the narration is what we call a "myth," and the charioteer a "mythological character." What this demonstrates is the essential role that myths play, in the narrating brain, in its arriving at the understanding of phenomena, and the extent to which mythological characters provide the actors on which the acquisition of not just behavior, but of "understanding," can operate. And so we come quite directly on the "gods" of mythology, which represent simply the actors necessary to animate phenomena of increasingly major extent, the ultimate god being the mythological character necessary to portray the behavior of the environment-as-a-whole, and so bring this ultimate "phenomenon" within the framework of "understanding."

It is important to see the absolutely essential role that *myths* play in the evolution of the understanding that characterizes the narrating brain, since there is a tendency to treat myths as somehow not based on anything "real," and so belonging to a trivial part of the fringes of human development and existence, of interest only to the most specialized of anthropologists. This is reinforced by the belief that, somehow or other, "science" has succeeded in providing humanity with a way of *escaping* this mythological basis for understanding the behavior of the phenomena that pervade the environment.

If this necessity for the existence of mythological characters, including gods, is combined with the previously explained potential for loss of selfness linked to malfunctioning of the narration processes of a brain, we can see that an illness related to such loss of selfness would include not just reports of control of behavior from "outside," but particularly control by mythological characters originating in narration, including control by *gods*.

It is important to distinguish between the evident necessity for the normal narrating brain to entertain the presence, "inside" it, of mythical characters, and the non-normal "feeling" in such a brain of *outside* control effected by such characters, and which represents a breakdown of "felt" selfness. We should also notice, by recalling the connection between narration and suggestibility, that it should be possible to instill in the normal narrating brain, precisely because it is normal, the "feeling" that it is controlled by mythical characters, but this control would then be felt as coming from *inside* the brain, and fall within the normal control of "felt" selfness.

Indeed, the brain affected in this way would be said to have been "taught" and "learned" to understand—even to "explain"—the portion of its environment represented in the myth acquired in this way. Such a brain would be viewed as holding-one-of-the- normal-"understandings"-and-"explanations"-of-the-way-the-environment-works.

A picture of the place of myths in a narrating brain allows us to situate, within it, what we know as a "mystery," since a "mystery" then becomes simply a phenomenon with which it is not possible for such a brain to associate any form of *even mythical* actors or narration. Thus, a "mystery" lies *outside* the boundary of understanding—of explanation—associated with a particular brain, and it can only be brought inside this boundary by behavior that we refer to as "discovery"; that is, by behavior that provides a myth that matches, at some level or other, the behavior embedded in the mystery.

Evidently, the myth must include such actors as might be necessary to support a narration, and is, in a significant sense, simply a copy of the phenomenon that constituted the mystery—a copy residing as an "*as if*"-it-*were*-the-phenomenon, in the brain of the hearer. A way of viewing a mystery is therefore as a phenomenon that it is not possible to map into a particular narrating brain, even by using all the capacity for myth-making and narration that it *then* possesses. Thus, a "discovery" will generally originate in a *single* brain as a local "mutation," and need to be taught to other brains, affected by the same mystery, by a process of narration.

There is no place, other than the "as ifs" of *neurons* in the brains of *human* hearers and speakers of grammar's *third* person, in which the myths and discoveries flowing out of narration could be born and grow.

Suppose, therefore, that the neuronal auto-destruction of neurodegenerative disease should attack those "as ifs" of *neurons* in the brain of a *hearer* in which the complex transfiguration of *phenomena*—presented as myths and their discoveries, accompanied by understanding and explanation—would normally take place.

Then we can expect an "illness" that manifests itself in the victim's incapacity to penetrate mystery by the fashioning of myths that "explain" experience. In the most extreme case, this would severely limit the capacity of narration to support the teaching and learning of behavior, and reduce the victim's "learned" behavior to the simple imitation of the directly observable behavior of other creatures, including, for instance, their stance, walk and vocalization. For this would then constitute the limit of the victim's understanding of them, that is, the limit beyond which all is mystery—as would have been the case for pre-narration creatures.

## NARRATION AND EMPATHY

If, as I am asserting, storytelling constituted evolution's last humanizing step, then it should be possible to find in this step the basis of the human feature that we identify as "empathy," since none of the evolutionary developments identified as occurring previously provide such a basis.

In the narrowest view that we can have of empathy, it is manifested in the capacity of one human to "identify with" the experiences of another; it is the basis of the ability to "feel for someone else." An example of this would flow from a question such as: How could an individual pre-narration humanoid creature come to know that dreaming, for instance, was not some kind of unique personal experience, but was experienced by others? Evidently, the answer is that such a creature could *not* come to know this, with all the fantastic behavioral ramifications that such an answer must imply. Indeed, only the later-developing capacity to narrate could serve to transform the hidden, individual experience of dreaming into a "shared experience," and so bring the derived waking behavior within the framework of understanding.

But in a broader sense, empathy is manifested in the capacity to identify with the "feelings" of other more general animate entities than just humans. Is there anything specific about narration that can be associated with the development of such a broadly based empathy? The answer is clearly yes, since the whole appearance of empathy in the human can be linked directly to the process of animation, which is an essential concomitant of narration, and to the *internalization* of the actors in narration that are a necessary part of it.

We can see some of the consequences of this by noticing that in the earlier creatures that preceded us, the act of *offensive* killing was linked to hunger. But we can observe quite easily in ourselves the fact that *offensive* killing is no longer linked instinctively to hunger—hunger doesn't drive us off to go and kill—and so the power of *offensive* killing which was, in pre-humans, under instinctive control, is, in us, under the control of learned behavior. Furthermore, this control is such that *offensive* killing, to the extent that it is indulged in at all, is "felt" as being "painful"—"as if" we can feel what must be the pain of the prey.

At least some of this control of learned behavior must be related to the learning associated with narration, and it is not difficult to see that much, if not all of this must come from animation and its expression in empathy. Such restraint on killing must come from the internalized pain of the actors in narration which induce in us some version of the pain, as narration conveys it, that would accompany their being killed. Empathy is among the clearest expressions of the "as ifs" of a brain's neurons—it is the expression of a narrating human being able to feel "as if" he or she shared the "feelings" of the *other* animated creature.

There is no place, other than the *neurons* in the brain of a narrating human host, in which the "feelings" of *other humans, especially,* as well as those of other creatures could come to be "as if" they were the host's "feelings"—no other place in which there could come to be the "feeling" of "empathy."

Suppose, therefore, that the neuronal auto-destruction of neurodegenerative disease should attack those "as ifs" of *neurons* in the brain of a *hearer* in which the "feelings" of *other humans, especially,* as well as those of other creatures could come to be "as if" they were the host's "feelings"—the feeling of empathy—would normally take place.

We could then expect to find some "illness" in which there would be, in its extreme form, disappearance of normal control from the human power to kill, leading to an illness in those so afflicted that would transform them into sources of danger to the humans and other creatures nearby, requiring their being restrained. But although the

*Keith C. M. Glegg*

extreme form of the illness would be manifested in killing, it could be manifested, much more generally, in cruelty of all sorts. The "illness" would be manifested in the behavior of the "sociopath."

## NARRATION AND UNTRUTH

Since a narration can refer to an "actor," in the most general sense, which *does not* have any existence *outside* the narration itself, a narration could present just such an "actor" *as if* it *did* have such *outside* existence. When a narration contains such a combination of "does not" and "as if," it is said to contain an "untruth" or "falsehood."

It is interesting to notice that the ability to generate narrations containing untruths is closely related to the ability to indulge in "make-believe," since, to get from the narration containing untruth to make-believe, all we need do is add to the narration containing untruth an indication that it *does* contain a (presumably "beneficial") untruth. Such indications take a variety of forms, all the way from the introductory "Once upon a time" of the spinner of fairy-tales, to the "Let us assume (make-believe) that" of the mathematician.

There is no place, other than the "as ifs" of *neurons* in the brains of *human* hearers and speakers of grammar's *third* person, in which the "as ifs" of *untruths* arising in narration could emerge and flourish.

Suppose, therefore, that the neuronal auto-destruction of neurodegenerative disease should attack those "as ifs" of *neurons* in the brain of a *hearer* in which the birth of the "as ifs" of narration's *untruths* would normally take place.

Then we can expect to find an "illness" in which, oddly enough, there is a *reduction* in the victim's production of *untruths*, simply because of a reduction in the ability to construct a narrative that can *convey* untruth—a *truth* "illness"! Such an illness could also lead to an inability to see make-believe as what it is—to see the world in "concrete" terms. One manifestation of this would be an inability to share the engaging near-untruthfulness of a Grimm or a Euclid.

## NARRATION AND UNREALITY

Prior to the evolutionary emergence of grammar's *third* person and, with it, narration, the utterances of an "I," in the *first* and the *second* person, related to what had actually been encountered by the "I" and a "You." However, that situation changes with narration, since all the actors conveyed in narration are simply phantoms that must be reconstituted by the hearer before it can make beneficial behavioral use of them, and some of them can even be the bases of untruths. Thus, we can distinguish between the actors present in the pre- and post-narration brain by saying that, while those in the earlier brain are all "real" actors, those in the later brain include animated and mythical actors, which can be called "unreal."

It seems reasonable, and I shall give evidence supporting this later that, as the narrating brain evolved, the components supporting narration occupied a later, separate, identifiable segment of the brain—as the narrating brain evolved, a "barrier" between the real and the unreal must have evolved with it. This would mean that there would be one "address" in the brain for real actors, and an evolutionarily later-occurring "address" for the phantom actors in narration.

There is no place, other than the *neurons* in the brains of *human* hearers and speakers of grammar's *third* person, in which evolution could have deposited the *separation* between the early "real" of "I"/"You," and the later "unreal" of narration.

Suppose, therefore, that the neuronal auto-destruction of neurodegenerative disease should attack those *neurons* in the brain in which the *separation* is realized.

Then we can expect to find an "illness" in which the actors from one of the "addresses" could migrate into that normally occupied by the others, and mix with them, so that the distinction between **real** and **unreal actors**, which comes easily from normally separate "addresses" in the one space or the other, would be lost. In this way, we can see that the neuronal auto-destruction of neurodegenerative disease could lead to the *inability* of the victim to *distinguish* between the *real* and the *unreal*.

However, the illness could go further, because all the animated actors of narration residing in a host's brain are *third* persons. Then, if that is so, *for the actors*, the *host* is *also* a *third* person—a reciprocal, mutual, *third* person. Thus, if the *host* should "hear" the *actors* speaking, they would speak about the **host,** all in the **third** person. This would not, in itself, be unusual or disturbing to the host, so long as the actors were originating in the space normally assigned to narration, and so were clearly distinguishable as unreal. However, in the presence of the kind of "illness" just mentioned, in which the host becomes unable to distinguish between the real and the unreal, the host will hear "real" actors "speaking" about him- or her-self, in the *third person*. And among these "real" actors will, on occasion, be found the gods of myths.

## A REMINDER

Having come such a long way from the evolutionary origin of neurodegenerative disease, it is easy to fall into seeing it, now, as detached from that origin. So, before going on to the discussion of schizophrenia it will be useful to have a brief reminder. The shortest way to begin is with recalling the following sequence:

> disorders in sleep → neurodegenerative disease → neuronal auto-erasure.

What this is reminding us of is that not only must a brain *sleep*, but, in case it doesn't, the consequences, indeed, the *evolutionarily determined* consequences, are the release of neurodegenerative disease.

And, this being so, it is useful to be reminded of what I called "the general answer" had to say about the full extent of neurodegenerative disease:

> Neurodegenerative disease involves the degenerative auto-destruction of the neuronal components that make up a brain. Therefore, as the degeneration progresses, **all the functions of a brain, whether "*sensorial*," or "*mental*," or "*directorial*," could eventually fail.** "Sensorial" functions are those associated with the brain's involvement in sensation, as in hearing, taste, touch, smell, and sight. "Mental" functions are those associated with brain-originated manifestations in the brain's host such as memory, imagination, "standing for," "as if," and so on. "Directorial" functions are those associated with the brain's direction of parts of its host: legs, hands, feet, tongue, for example.

So, it is not surprising that *all* the foregoing cases of "mental illness," as will be the case with schizophrenia, are, in one way or another, expressions of the enormous ramifications of neurodegenerative disease, with the most fundamental evolutionary source of its release being *disorders in sleep*.

It will also be noticed that, in the symptoms of schizophrenia that I'll be giving, there is considerable concern with whether it is or isn't a hereditary illness. So, it will be useful to be reminded of what was said earlier about the hereditary nature of *sleep disorders*, since they release neurodegenerative disease, and—as I would now have it—schizophrenia is simply another of the many faces of neurodegenerative disease. Here then is what was said of sleep disorders and heredity:

> [S]leep emerged as a *hereditary* feature of its host, expressing the host's augmented *genetic* endowment. But if sleep is such a feature, then there must be **genetic disorders** that are expressed in **sleep disorders**—there must be at least *some* sleep disorders that are *hereditary*. In effect, there must be some sleepers who suffer sleep disorders simply because one or both parents suffered sleep disorders, before them.
>
> However, we can go further. For, clearly, not *all* sleep disorders will be expressions of *genetic* disorders, since all sorts of disturbances, external and internal to a brain, can cause sleep disorders—even just writing a book on neurodegenerative disease can cause sleep disorders! In effect, sleep disorders can be either *hereditary* or *accidental*.
>
> Furthermore, there are other kinds of accident, other than accidental sleep disorders, which have an effect in a brain *similar* to that of a sleep disorder, because they are traumatic enough to produce "unclearable errors" in its configuration store. Such accidents include stroke, drug intake, direct head injury, encephalitis, concussions as in some sports, as examples.
>
> As we will see, these sleep disorders, hereditary or accidental, will have a lot to do with the onset of

neurodegenerative diseases, and whether *they* are, or are not, hereditary. And, except in some special situations, I'll refer to them all as just "sleep disorders."

It should therefore not be surprising to find, expressed in the symptoms of schizophrenia to be given, the concern with whether the illness is or is not hereditary—is or is not "genetic" in origin—since, as the foregoing passage is implying, it will, in some cases, be hereditary, and, in some cases, not.

## NARRATION AND SCHIZOPHRENIA

Given the relatively late-arriving evolutionary endowments that must underlie the emergence of narration in humans, we can reasonably assume, and I shall confirm this with experimental evidence later, that there exists some identifiable, evolutionarily-recent segment of the human brain that is responsible for narration. I shall refer to this segment of the brain as the "narration complex," and assume that it is responsible for the functions associated with the links between grammar's *third* person with its actors needing animation, and the older parts of the brain connected with "real" actors, and the use of grammar's *first* and *second* persons only.

Thus, the narration complex will be responsible for a long list of human phenomena and functions of the type discussed earlier. These include narration itself and the *third*-person components of grammar; the whole set of relations connected with empathy, which include the capacity to "feel for others"; the phenomenon of suggestibility and the possibility of being led and hypnotized by means of narration; the ability to animate and copy the behavior of non-human entities, and so, using myths, come to "understand" them; the ability to construct untruths, including those offered by a Grimm or a Euclid, with their introductory alerting excuses; the presence in the brain of the unreal, and its normal separateness from the real.

All these, arresting in their humanness and range, come with the narration complex when it functions normally, as evolution and natural selection have determined "normal" function. But, as shown earlier, these symbols of humanness can be undone by the neuronal auto-destruction of neurodegenerative disease, giving rise to a wide range of "illnesses," which, as I am now advancing, are the complex, and disparate departures from the normal that are witnessed in the group of behavioral abnormalities that go to make up *schizophrenia*. In support of this view, I shall quote some of the more prominent of the numerous observed symptoms of schizophrenia, so that they can be compared with the behavior expected from failures of the narration complex discussed earlier.

The symptoms of schizophrenia have been studied extensively, and described by many people in many different ways, at widely different times. I have selected passages from two of these descriptions, and the first selection, from *Brain, Mind and Behavior*, by Bloom, Lazerson and Hofstadter, in which all the emphases have been added by me, is as follows:

> "Converging lines of research have led to the modern view that schizophrenic diseases have a *biological* basis . . . However, before we discuss this evidence in detail, it is proper to point out that no specific causes of schizophrenia have yet been directly identified." (Page 265)

> "The genetics of schizophrenia are relatively complicated, but still speak a rather clear message. Some *inheritable* predisposing 'factor' can lead to the development of schizophrenia." (Page 267)

> "The overall behavior of schizophrenic patients is primarily characterized by abnormally distorted perceptions of *what is real and what is not.* Some patients *hear voices.* . . . They believe that their ideas of the world are imposed on their minds by *outside* forces, and thus overcome, they are unable to separate fact from fancy. . . . At the same time, schizophrenic patients are *unable to generalize.* . . . This inability to make generalized abstractions is said to arise from an extremely *'concrete'* way of looking at the world. . . . Still others show periods of *extremely disruptive aggressive behavior* and require restraint to avoid *hurting themselves and others.* It seems clear that a disturbance of the thinking process occurs in all of these patients.

But it is not at all clear that the same, unknown cause is the source of the problem in all cases, or that these extremely varied clinical problems can have the same biological basis. . . . Psychiatrists long found it difficult to understand the relationship between the 'positive symptoms' that are suffered by some schizophrenics – hallucination, thought disorders, and *delusions* – and the 'negative symptoms' expressed by others – loss of emotional responses, *inanimate postures, loss of spontaneous speech* and general lack of motivation." (Pages 260-261)

The second selection, which is from *Behavioral Neurology*, by Pincus and Tucker, and in which the emphases are again mine, is as follows:

"1. Auditory Hallucinations
   a. Audible thoughts (*voices speaking patient's* thoughts aloud)
   b. Voices arguing (two or more voices arguing *usually about patient – refer to patient in third person*)
   c. Voices commenting on *patient's* actions
2. Delusional Experiences
   a. Bodily sensations imposed on patient by some *external* source
   b. Thought being taken from his mind
   c. Thoughts ascribed to *others*
   d. Diffusion of thoughts (patient's thoughts experienced as *all around him*)
   e. Feelings, impulses, volitional acts imposed on him or under the control of *external* sources . . . . "
   (Page 62)

Pincus and Tucker discuss the genetic question at considerable length, and, after citing a number of studies, conclude:

"These studies do more than merely offer support for a theory about genetic influence in schizophrenia. They indicate that *it is a genetic disease*. They offer no support for the view that psycho-social environment

plays any role in determining the risk of developing schizophrenia in individuals who are generally at high risk. The child of a schizophrenic has the same chance of developing the disease whether he is raised by his schizophrenic parent or in a normal environment."
(Page 89)

In view of the quite substantial correspondence between the symptoms that one can predict by using the assumption of a narration complex and failures in it, and the symptoms actually observed in schizophrenia, it seems reasonable to conclude that the narration complex must exist as a separate, identifiable segment of the human brain, and that schizophrenia is brought on by departures from normal biological function that are induced in this complex by the neuronal auto-destruction of neurodegenerative disease, giving rise to a wide range of disparate "illnesses," as explained earlier.

This leads to two other aspects of schizophrenia—age at onset, and autism—that are conveniently introduced by the following additional quotes from *Behavioral Neurology* in which the emphases are again mine:

"Schizophrenia is primarily a disease of *young people.* Kraepellin noted that most patients were *under the age of thirty-five years* at the time of diagnosis, a finding which has been confirmed in many subsequent studies. The first clear-cut symptoms appear *before the age of twenty-five* in 50 percent of the cases; onset after the age of forty is unusual (Kraepelin, 1925). Symptoms rarely begin in the first decade, but when they do, they virtually always occur in the latter half, never before the age of five. *Childhood schizophrenia* has often been *confused* with *infantile autism*, a condition which usually begins in the first year of life and *always appears before age five* . . . ." (Page 60)

"*Disorders of speech* are the hallmark of early childhood autism, a behavioral syndrome first described by Kanner and *often mislabeled* 'childhood schizophrenia.' . . . It begins in the first few years of life. . . . Autistic

113

children have a marked inability to form human relationships and give a sharp impression of extreme solitariness. . . . The most convincing evidence that the two conditions [autism and schizophrenia] are not identical is *genetic. There is no increase in the prevalence of schizophrenia in the parents or siblings of autistic children.* . . . Autistic children often have varying deficits in comprehension, symbolic thinking, and the formation of abstract concepts." (Pages 124-5-6)

## AGE AT ONSET

Going to the beginning of the first of these two quotes, it is interesting to notice the difference that an evolutionary treatment of the origin of schizophrenia can bring to the view that it "is primarily a disease of young people [because] the first clear-cut symptoms appear before the age of twenty-five in 50 percent of the cases; [and] onset after the age of forty is unusual." For, if we imagine the situation of the first humans in which a narration complex had emerged, and who were then able to suffer from schizophrenia, the average life-span was almost certainly no more than about twenty years, and so schizophrenia, which would show its symptoms at age twenty-five, would have been very much a disease of *old age.*

Thus, it could be that, in this disease of some of the "young people" of today, we can see a dim evolutionary marker of the approaching end of what must have been, at best, the relatively short life of our nearest evolutionary ancestor. It was doubtless then, as it is now, that *old age* was linked to increases in *disorders in sleep*, with a consequent increase in neurodegenerative disease, and *all* its expressions, of which schizophrenia is only one.

## AUTISM

We can come on another of these age-related aspects of schizophrenia by noticing that if, as I believe should now be clear, the illness is a manifestation of inadequacies of one sort or another in the functioning of the brain's narration complex, then we might expect to find what I would call "normal, developmental schizophrenia" in *young children,* during the period when they are capable of short, single-sentence speech *only*, mainly grammar's *first* and *second* persons, and which I take to be an indication of the *incomplete development* of their *narration complex*. Evidently, such a *normal* manifestation of incomplete development of the brain of the young child would *gradually disappear in the course of normal maturation of the brain*—the *normal* childhood development of the narration complex comes *relatively late* because its *evolutionary arrival* was a *relatively late* addition to a brain.

I therefore believe that much of what are regarded as episodes of "childish behavior"—including delusions, the inability to cope with complex abstractions, episodes of cruelty, long series of "why" reflecting an inability to penetrate mystery—are simply manifestations of normal, developmental schizophrenia, *which ultimately disappear in the normal child, as its brain, and the narration complex with it, become fully mature.*

And it needs to be stressed that such developmental schizophrenia is hereditary only in the almost trivial sense that *all* normal biological development is hereditary, and so one would not expect to be able to find any particular evidence of *schizophrenia* among parents or siblings, linked to such normal development of almost every child.

However, the course of *normal* development of the narration complex might be *arrested* for a variety of purely *accidental* reasons, such as brain-damage suffered at birth, or simply a blow on the head, which have nothing to do with genetics and heredity. When normal development of the narration complex is arrested in some such non-hereditary way, the child so stricken will then be shackled to a condition of *permanent* schizophrenia, which displays almost all the symptoms of the schizophrenia of "old age," but which has *no connection* to parental schizophrenia. And I add "almost," since such schizophrenia in children will lack the great mass of narration-based experience and "learning" on which the illness can draw in adults, and which will tend to give the symptoms in adults a broader and even more overtly perplexing quality.

I believe that this kind of *accidentally* induced *permanence*, in what is normally the *passing, developmental schizophrenia of childhood*, gives rise to the illness referred to in the two foregoing quotes as *"autism."* If you read them once more—note, in particular, the sentence: *"There is no increase in the prevalence of schizophrenia in the parents or siblings of autistic children*"—you will see how closely such an explanation fits their facts, as well as how difficult it would have been, in the absence of an evolutionary sketch of the development and functions of the narration complex, to avoid the confusion in interpretation which is now evident in them. "Childhood schizophrenia" is not at all a "mislabel" for autism; on the contrary, this label reflects the well-developed, collective intuition of the observant clinicians who simply tried to put their experience into words.

Since the evolutionary emergence of the narration complex must have been gradual, the maturation of the narration complex must take place gradually over the course of normal childhood development. There must, therefore, be "degrees" of autism, depending on where or perhaps when, along the course of normal development, accidental arrest of development has taken place. Such degrees of autism are well known—from the most severe and evident, to the almost un-diagnosable.

Viewed in this light, it is easy to understand the anguish inspired by the more severely affected autistic children, growing into adults, having been accidentally burdened with much of the "unreachable," that is typical of adult schizophrenia. Enter the caregiver.

## SUMMARY

This has now become such a long discussion of schizophrenia that it seems desirable to present a summary sketch of the implications of what the present chapter has had to say. The picture I arrive at is one in which the creature that preceded us, as the storytelling humans that we became, must have spoken single sentences, limited to grammar's *first* and *second* persons only. Lacking the capacity to narrate, with all that this can now be seen to imply, the behavior of the creature must have been somewhat like that of the human child, still displaying *normal* developmental schizophrenia.

The survival benefits of narration are so great that the behavior determined by the associated complex slowly came to dominate more and more of the entire behavior of the creature. The process must have been slow, because of the enormous increase in memory that is required to cope with the streams of second-hand "experience" associated with narration. As the creature continued to evolve, it continued to display developmental schizophrenia in its children, but the normal course of development came to be that in which the power of narration captured the behavior of the mature creature. Thus, the mature period of its life was spent in the exploitation of the benefits of narration and its associated powers.

It seems reasonable that the brain of this early "human" would not have been able to support the immense demands of narration for more and more memory. Thus, as a normal feature of its old age, neurodegenerative disease, released by increasing disorders in sleep, became dominant. However, the combined effects of further slow mutation, and the power of natural selection, linked to narration itself, gradually extended the capacity of the aging "human" brain to produce increased memory. In this way, a line of humans developed who were able to survive longer, without suffering the previously normal advance of neurodegenerative disease's disabilities, *including schizophrenia*, leaving other afflictions to usher in the ends of their lengthening lives.

But traces of the earlier "human" condition rest in the genes of the general human population, so that, today, some fraction of it remains vulnerable to relapse into schizophrenia at a time that seems now to be *late youth*, but which was really *old age* at the time of the emergence of narration. Given all this, it would be reassuring if some evolutionarily late-arriving place could be identified in the human brain, with which the narration complex could be identified.

## EVIDENCE FOR EXISTENCE OF A NARRATION COMPLEX

As you will see, the evidence I'm going to be citing involves the work of Dr. Paul D. Maclean, on the brain's evolutionary development. And, although what Dr. Maclean saw as the "triune brain" has received considerable criticism, it serves as a simple and adequate picture for what I'm intending to show.

So I'll now go to *The Brain*, by Restak, to show that there is a place in the human brain that can be associated with the narration complex. As usual, the emphases are mine in the following:

> The third 'brain', the cortex is most highly developed in humans. It is a kind of problem-solving and memorizing device to aid the two older formations of the brain in the struggle for survival. Dr. MacLean

compares the cortex to a 'computer' that can look into the future and anticipate the consequences of actions. The cerebral cortex furnishes us with our most human qualities: our language, our ability to reason, to deal with symbols, and to develop a culture. The *prefrontal* areas of the cortex are the most highly developed. Dr. MacLean considers the development of the *prefrontal fibres* the most auspicious turn of events in the history of biology. 'It is this new development that makes possible the insight required to plan for the *needs of others* as well as the self, and to use our knowledge to *alleviate suffering* everywhere. In creating for the first time a creature with a *concern for all living things*, nature accomplished a one-hundred-eighty-degree turnabout from what had previously been a reptile-eat-reptile and dog-eat-dog world.' . . ." (Restak, pages 136-7)

Evidently, the place in the brain that I am looking for is known as the "prefrontal areas of the cortex." This is clear from the references by Dr. MacLean to a number of modes of behavior which express *empathy*, and which therefore link to animation and hence narration. It is also significant that the prefrontal areas are identified as a "new development." Support for this conclusion, of a quite different kind, comes from recent comparisons, increasing in number, of brain activity in normal and schizophrenic individuals. These comparisons were made using scanning devices capable of identifying the level of chemical activity in parts of the functioning brain. What is found is that schizophrenics show very clear deficiencies of chemical activity in the prefrontal cortex.

We can see from this that the new developments leading to the narration complex must have led to a new species, since the changes that evidently took place in the prefrontal cortex are too extensive to have allowed successful mating between those who had come before and those who came after the changes. And the new species must be us, since the changes account for the last significant development in our behavior.

It is also clear that, at the time of our final emergence as a single species, we were starting to tell stories. Consequently, we can begin to see the possibility of the stories that were handed down from generation to generation having among them some that recount, maybe dimly, the story of this last upheaval which produced the animal species that is ours.

There must be, among these stories, some that tell about the disappearance of every living trace of the creatures just one step back, and which, at the start, were all around the first few of us; stories that tell about the ancient ones who could make a sentence, and learn from their parents. They even dreamed dreams, but, and here's the catch, couldn't share them; nor could they take suggestions.

Seems they paid a high price for just being themselves, and minding their own "I"-"You" business. But it's not taking our suggestions, I guess, that really did them in, for our ancestors, even back then, must surely have tried suggesting ways for them to be *slaves*, thereby avoiding extinction. And maybe they even *did* try being slaves, but when they came to really understand what that meant, with the befuddling mystery of *third*-person chatter about ghosts all around, they opted for extinction as clearly a better option.

## THE EVOLVING HUMAN

Where there is room for possible biological specialization, evolution will explore the room using living entities as the vehicles of exploration, and their survival or extinction as a measure of the amount of room available. So having arrived at the human species, evolution continued to explore the possibilities of specialization. What, then, are the possibilities?

In the first place, there is the entire biological system that preceded the emergence of the human species, and on which our basic existence is founded. Thus, as humans moved about, exposing themselves to new physical environments, minor evolutionary changes could occur beneficially in those parts of this basic biological system that interacted most directly with the various physical environments. Evidently, such changes did occur since, although humans belong to the same species, and must have originated in a small homogeneous group, we presently embody numerous physical variations, as are readily apparent in the differences—in skin-color and hair-type, for instance—to be found among us. This sort of physical change, in response to changes in physical environment, was certainly a way in which evolution could, and, evidently, did do its specializing.

There is also a second way in which specialization can proceed: by acting on the narration complex, as the last evolutionary transition leading to humans. Because this last transition is also embedded in biology, and, as surely as it was established by evolution, it can continue to be moulded and specialized by it.

But evolutionary changes here would differ quite markedly from those that affect mainly external, physical characteristics, because the narration complex is the source of every mode of *behavior* that we regard as being uniquely human. This includes everything from our ability to construct theories, to schizophrenia; from our ability to tell stories, to our willingness to accept others, especially different others, into our lives. So it would take a larger book than this to explore what even minor differences in the evolutionary specialization of the narration complex could have meant, for hereditary differences, in the human-specific behaviors of the branches of our species, which wandered off to separate corners of the earth.

Of course, I know how anxiously racists of every "race" await confirmation, and how happily some of them will seize on this last paragraph. Let me say, therefore, that I see no reason for pretending that what seems to have been a reasonable set of possible evolutionary specializations could not have occurred, or did not occur.

However, I suggest to the more impatient racists that they wait for a fuller exploration of the possibilities—by Evolution *herself,* for instance. She comes well recommended, never having committed an error, thanks to her faithful companion Natural Selection. She knows where the fossils are buried, especially those of the *great masters* of their times and places, still known to her alone. So now she's too shrewd to place her bets . . . before the game's a little further gone—and Natural Selection has blown another whistle.

# CHAPTER IX
# ALMOST LIKE SLEEP

## THE TANDEM-BIKE RIDER

There's this truly marvelous story of an unintended outcome, following the action of a devoted caregiver. A year or so ago, a patient suffering from Parkinson's disease was taken on a ride as the back-rider on a tandem bicycle, at relatively high speed, for a few hours. That's what devoted caregivers are like—without them …. On dismounting, the patient's **symptoms**, but not only those that would be associated directly with pedaling, **decreased substantially**, and remained so for days.

In the absence of the present evolutionary explanation, the tendency might be to attribute the result to still another of the well-known benefits attendant on "vigorous exercise." However the present evolutionary explanation suggests a different way to account for this remarkable result, bordering as it does, on the miraculous.

To see what this might be, I'll begin with bringing together three reminders—"A," "B" and "C"—of earlier implications or conclusions.

Reminder (A):
This reminder concerns the behavior that characterizes Parkinson's disease, as follows:

> The *un-control* in the behavior of motor extremities, which characterizes behavior in a victim of *Parkinson's disease*, is an expression of the ongoing failure of the brain to direct the victim's motor extremities into a ***new configuration*** that would ***follow directly*** from the

**actual existing** *configuration* of the motor extremities themselves. The simplest evidence of this condition is imprecision in the brain's placing of, and trembling in a motor extremity. This failure is due to **errors** in the *representation,* in the victim's *configuration store* (motor homunculus), of the *actual existing* configuration of the victim's motor extremities. The **errors** will have been accumulated following persistent **disorders in sleep** (or some event that will have had an equivalent effect), leading to the *neuronal auto-destruction* of neurodegenerative disease.

Reminder (B):
This reminder, form Chapter IV, concerns the way in which **sleep** allows the **errors** in the *configuration store* to be **cleared**. And it's so long that, in addition to the usual inset of margins, I've added ↓ and ↑ at the start and finish. The reminder is as follows:

> ↓ Picture, then, the behavior of a host in which the directing of its parts is being carried out by this new kind of brain [the one that's going to include sleep, as soon becomes evident]. And, in this brain, as in every brain, errors accumulate in its configuration store. As the errors accumulate, the brain-directed behavior begins to endanger the host's survival. But the ensuing **stress** in the host **does not** lead to the brain destroying itself, *as would have occurred with the previous one-shot brain.* **Instead**, the *augmented genetic development* of the host is now expressed in the *stress* releasing a burst of brain-substance ... which induces a state of **quiescence** in the host—its motor parts become **still**.

During this state of **quiescence**, the configuration store is **not** being continually refreshed with **new, changing** configurations from the host's **now stilled** motor parts. Instead, the configuration store is forced to **repeatedly** acquire the **same stationary, resting configuration**— *w*hatever it happens to be—and becomes reloaded with **that same stationary** configuration, *over and over* again.

And now I recall an earlier conclusion, reached in Chapter II, regarding error-correcting codes:

> An *error-correcting code can be constructed*, allowing arrival at a *message "almost" free of errors, by repeating* the *same* message with its errors *over and over*, and using, as the *final* message, *the version obtained by taking together all the parts that keep on repeating.* And this would apply to the parts of an *image*, if the parts were sent as a message.

Returning attention to the configuration store, it continues to **repeatedly** acquire the **same stationary, resting configuration**, whatever it happens to be, *while applying an error-correcting code*, based simply on *repetition—retaining* in the configuration store only those parts of the incoming image that **repeat**.

Then, slowly, the effect of the burst of quiescence-inducing brain-substance wears off, and the host recovers its active behavior, with its *configuration store* initially "almost" **free of errors.**

This new kind of *error-correcting* brain can now resume the directing of its host's parts. And it can do this, *time after time*, as its stress-induced error-clearing periods of quiescence come and go.

> ... Although, in this way, the host's *augmented genetic development* would have succeeded in clearing the errors from a brain's configuration store, it is important to notice that it does so at the expense of what would be an **endless series of periods of quiescence** which interrupt the host's full functioning. These periods of quiescence, which characterize the functioning of a host with a brain having the new ability to clear the errors in its configuration store, are the earliest evolutionary manifestations of what we come to know as "**sleep.**" When the creature is no longer in one of these periods, we say it is "awake," by which we evidently mean: functioning with its neuron-based brain doing its directing, with a configuration store operating as it did earlier—initially free of errors.

It is important to notice that there is no particular configuration-of-motor-extremities to which the configuration store must be returned during sleep—there is no "zero" configuration of motor extremities from which the waking state must always begin. The waking state begins with whatever essentially "fixed" configuration of motor extremities happens to have prevailed during the *stillness* of sleep, and goes on from there, by continued refreshing of the configuration store, with whatever stress-inducing errors accumulate in the course of refreshing. The stress induced by these accumulating errors, while awake, brings on the next period of sleep, in an endless series of sleep and waking.↑

Reminder (C):
This reminder, also from Chapter IV, is about the repetitions necessary, during sleep, to clear errors from a brain's configuration store. It is as follows:

> It takes something like **twenty minutes**, and 20,000 repetitions, coming from a host's motor parts, for the errors in the configuration store of a brain like ours to be returned to a level that allows renewed rapid, precise, stress-free control of our motor parts. Presumably, that many repetitions are required because of the enormous size of the "messages" involved in making even a single "image" of the host's motor parts in the configuration store.

Now, so that we can place the patient in the context of these Reminders, let us notice the following conditions affecting a tandem back-rider:

1.  On a tandem bike, the back-rider must, to a considerable extent, "give-up-brain-control" to the front-rider—in much the same way that, **during sleep**, the sleeper must "give-up-brain-control" during a state of "quiescence." And although the back-rider does *not* fall into a state of quiescence, the extent to which a back-rider must "give-up-brain-control" is here taken to "approximate" the state of quiescence in sleep. This attaches the back-rider to Reminder (B).

2. Because of the way the tandem bicycle is built, the back-rider is forced to take on the ***same configuration*** time after time, after time.... So, although the *configuration* of the back-rider's motor-extremities is not **still**, as in sleep, it **repeats** with relatively high accuracy the ***same configuration***—almost "as if" it were **still**. This also attaches the back-rider to Reminder (B).

3. If I make the assumption that there would have been something like a hundred (100) cycles of pedaling per minute, then that would give about twelve thousand (12,000) cycles in a two-hour ride. Notice that this is *of the same order* as the "20,000 repetitions," in Reminder (C), that are necessary for *clearing of errors* from the configuration store.

Even with just this much, you have probably already guessed where my explanation is leading. Yes, what I'm going to give as the explanation—of the almost miraculous result in which, after a few hours as the back-rider on the tandem bike, the patient's symptoms of Parkinson's abated for a while—is as follows:

> The conditions affecting the patient, given in conditions (1), (2) and (3) above, provide an **approximation** to the *conditions in **sleep**,* as in Reminder (B), which normally allow the clearing of errors from the configurations store. The **approximation** to *conditions in **sleep*** is *close enough* for the back-rider on the tandem bike to have *substantially cleared the **errors** from the **configuration store**,* thereby freeing the patient from the symptoms of Parkinson's, as these are given in Reminder (A). Since the symptoms reduced are *not limited* to those that would be associated with just arms, legs and pedaling, this tends to confirm that the *affected* "region" in the configuration-store is, indeed, *all of it*, as would have been the case in *actual* sleep, whatever the external motor configuration had then been.

However, errors in the configuration store will accumulate once more, as disorders in sleep are brought on by the continuing neuronal auto-destruction of the very neurodegenerative disease itself, of which Parkinson's is one symptom. And this will render relief from the symptoms of the disease temporary.

All this evidently raises an extremely important *general* question regarding a possible way of relieving, if only temporarily, the symptoms of at least some neurodegenerative diseases, by not just tandem-bike riding, alone. The question is as follows:

> Can the symptoms of a neurodegenerative disease *generally* be reduced by the patient **repeating**, *some many thousands of times*, exercises involving *motor-extremities*, while conditions *approximating the brain's condition in sleep—giving-up-control—have been established*?

The answer to this question is evidently one that should occupy researchers in neurology, rather than more lines here. Clearly, however, the greatest challenge resides in how "conditions *approximating the brain's condition in sleep—giving-up-control*" would be established, especially in patients already suffering the effects of neurodegenerative disease. And this serves to highlight the peculiarly favorable—albeit evanescent—condition that was presented by the back-rider on a tandem bike.

Not without sorrow, for the caregiver.

# EPILOGUE

The words of the Preacher, the son of David, king in Jerusalem.

Vanity of vanities, saith the Preacher, vanity of vanities; all is vanity.

What profit hath a man of all his labor which he taketh under the sun?

One generation passeth away, and another generation cometh: but the earth abideth for ever.

The sun also ariseth, and the sun goeth down, and hasteth to his place where he arose.

The wind goeth toward the south, and turneth about unto the north; it whirleth about continually, and the wind returneth again according to his circuits.

All the rivers run into the sea; yet the sea is not full; unto the place from whence the rivers come, thither they return again.

All things are full of labor; man cannot utter it: the eye is not satisfied with seeing, nor the ear filled with hearing.

The thing that hath been, it is that which shall be; and that which is done is that which shall be done; and there is no new thing under the sun.

Is there any thing whereof it may be said, See, this is new? It hath been already of old time, which was before us.

There is no remembrance of former things; neither shall there be any remembrance of things that are to come with those that shall come after.

I the Preacher was king over Israel in Jerusalem.

And I gave my heart to seek and search out by wisdom concerning all things that are done under heaven; this sore travail hath God given to the sons of man to be exercised therewith.

I have seen all the works that are done under the sun; and, behold, all is vanity and vexation of spirit.

That which is crooked cannot be made straight: and that which is wanting cannot be numbered.

I communed with mine own heart, saying, Lo, I am come to great estate, and have gotten more wisdom than all they that have been before me in Jerusalem: yea, my heart had great experience of wisdom and knowledge.

And I gave my heart to know wisdom, and to know madness and folly: I perceived that this also is vexation of spirit.

For in much wisdom is much grief: and he that increaseth knowledge increaseth sorrow.

ECCLESIASTES: Chapter I

# REFERENCES

Bloom, Floyd E.; Lazerson, Arlyne; Hofstadter, Laura, 1985, *Brain, Mind and Behavior,* New York, W. H. Freeman and Company.

Darwin, Charles, 1859, *The Origin of Species.* Since 1958 a reprint has been available as A Mentor Book, Times Mirror, New York.

Jamison, Kay Redfield, 1999, *Night Falls Fast,* New York, Alfred A. Knopf.

MacLean, Paul D., *Triune Brain,* Encyclopedia of **Neuroscience,** Boston, Volume II, 198?, edited by George Adelman, pages 1235-37, BIRKHAUSER.

Pincus, Jonathan H.; Tucker, Gary J.; 1974, *Behavioral Neurology,* London, Oxford University Press.

Restak, Richard, 1984, *The Brain,* Toronto, Bantam Books.

www.ingramcontent.com/pod-product-compliance
Lightning Source LLC
Chambersburg PA
CBHW021954170526
45157CB00003B/992